Mathematics Study Resources

Volume 5

Series Editors

Kolja Knauer, Departament de Matemàtiques Informàtic, Universitat de Barcelona, Barcelona, Barcelona, Spain

Elijah Liflyand, Department of Mathematics, Bar-Ilan University, Ramat-Gan, Israel

This series comprises direct translations of successful foreign language titles, especially from the German language.

Powered by advances in automated translation, these books draw on global teaching excellence to provide students and lecturers with diverse materials for teaching and study.

Oliver Stein

Basic Concepts of Global Optimization

 Springer

Oliver Stein
Institut für Operations Research (IOR)
Karlsruher Institut für Technologie (KIT)
Karlsruhe, Germany

ISSN 2731-3824　　　　　　ISSN 2731-3832　(electronic)
Mathematics Study Resources
ISBN 978-3-662-66239-7　　ISBN 978-3-662-66240-3　(eBook)
https://doi.org/10.1007/978-3-662-66240-3

This book is a translation of the original German edition "Grundzüge der Globalen Optimierung", 2nd edition, by Oliver Stein, published by Springer-Verlag GmbH, DE in 2021. The translation was done with the help of an artificial intelligence machine translation tool. A subsequent human revision was done primarily in terms of content, so that the book will read stylistically differently from a conventional translation. Springer Nature works continuously to further the development of tools for the production of books and on the related technologies to support the authors.

Translation from the German language edition: "Grundzüge der Globalen Optimierung" by Oliver Stein, © Springer-Verlag GmbH Deutschland, ein Teil von Springer Nature 2021. Published by Springer-Verlag GmbH. All Rights Reserved.

© The Editor(s) (if applicable) and The Author(s), under exclusive license to Springer-Verlag GmbH, DE, part of Springer Nature 2024

This work is subject to copyright. All rights are solely and exclusively licensed by the Publisher, whether the whole or part of the material is concerned, specifically the rights of translation, reprinting, reuse of illustrations, recitation, broadcasting, reproduction on microfilms or in any other physical way, and transmission or information storage and retrieval, electronic adaptation, computer software, or by similar or dissimilar methodology now known or hereafter developed.
The use of general descriptive names, registered names, trademarks, service marks, etc. in this publication does not imply, even in the absence of a specific statement, that such names are exempt from the relevant protective laws and regulations and therefore free for general use.
The publisher, the authors and the editors are safe to assume that the advice and information in this book are believed to be true and accurate at the date of publication. Neither the publisher nor the authors or the editors give a warranty, expressed or implied, with respect to the material contained herein or for any errors or omissions that may have been made. The publisher remains neutral with regard to jurisdictional claims in published maps and institutional affiliations.

This Springer imprint is published by the registered company Springer-Verlag GmbH, DE, part of Springer Nature.
The registered company address is: Heidelberger Platz 3, 14197 Berlin, Germany

If disposing of this product, please recycle the paper.

*Don't panic.
(Douglas Adams)*

Preface

This textbook originated from the notes of my lecture "Global Optimization I and II", which I have been giving annually at the Karlsruhe Institute of Technology since 2008. The primary audience of this lecture are students of Industrial Engineering and Management in the Bachelor's specialization program. In the present textbook, this is reflected in that mathematical facts are treated stringently, but are significantly more motivated and illustrated than in a textbook for a purely mathematical course. The book is therefore aimed at students who want to understand and apply mathematically sound methods in their course of study, as is the case in the natural, engineering, and economic sciences. Since a more detailed motivation naturally comes at the expense of the scope of the material, this book limits itself to the presentation of the *basic concepts* of global optimization.

The subject is the treatment of global minimization or maximization models with nonlinear objective functions under nonlinear constraints, as they often occur in application disciplines. The problem often arises that numerical solution methods can efficiently find *local* optimal points, while *global* optimal points are much harder to identify. This corresponds to the fact that with local search methods one can easily find the peak of the nearest mountain, but the search for the peak of Mount Everest is rather complex.

Before turning to the theoretical and algorithmic identification of global optimal points, however, it is important to clarify whether an optimization problem possesses optimal points in the first place. The introductory Chap. 1 therefore goes into detail on the different types of unsolvability as well as on mild sufficient conditions for solvability.

Chapter 2 discusses how to determine global minimal points of smooth convex optimization problems, because not only strong theoretical results can be given for them but also efficient algorithms. A central result for convex optimization problems is that every local minimal point must necessarily also be a global minimal point. Therefore, methods of local optimization [37] are sufficient for the global solution of convex optimization problems. Since it is "simple" in this sense, convex optimization is often not assigned to the field of global optimization, which then rather focuses on the solution of "difficult" nonconvex optimization problems. However, this textbook uses techniques of convex optimization from Chap. 2 to solve nonconvex problems in Chap. 3, which explains the chosen structure. In

Chap. 3, a branch-and-bound method based on interval arithmetic is derived as an exemplary and practically implementable method of global optimization.

This textbook can serve as the basis for a four-hour lecture. It partly relies on presentations by the authors M.S. Bazaraa et al. [4], S. Boyd and L. Vandenberghe [5], C.A. Floudas [10], O. Güler [14], J.-B. Hiriart-Urruty and C. Lemaréchal [19], R. Horst and H. Tuy [21] as well as H.Th. Jongen et al. [24], who also deal with many questions beyond this book. For basics of (local) nonlinear optimization, see [37], and for general basics of optimization, see [30].

The present textbook does not aim at giving a comprehensive literature review. In particular, since it is based on an automated translation of the German original textbook version, several other German books are cited, partly authored by myself. For the latter, please check for translations generated after the publication of the present book.

At this point, I would like to thank the Springer staff for their very helpful cooperation in translating the manuscript and copy editing.

A big thank you also goes to my current and former PhD students Maren Beck, Dr. Tomáš Bajbar, Prof. Dr. Peter Kirst, Dr. Robert Mohr, Dr. Christoph Neumann, Dr. Marcel Sinske, Dr. Paul Steuermann, and Prof. Dr. Nathan Sudermann-Merx as well as to numerous students who have pointed out to me possibilities for content and formal improvements during the development of this teaching material. The present text was typeset in LaTeX2e. The figures come from *Xfig* or were generated as output from *Matlab*.

Text set in smaller typeface indicates material that is provided for completeness, but can be skipped on first reading.

Karlsruhe, Germany
May 2024

Oliver Stein

Contents

1 **Introduction** 1
   1.1 Examples and Terminology 2
   1.2 Solvability 12
      1.2.1 Definition of Solvability 12
      1.2.2 Types of Unsolvability 14
      1.2.3 The Weierstrass Theorem 18
      1.2.4 Unbounded Feasible Sets 20
      1.2.5 Nonclosed Feasible Sets 28
   1.3 Rules of Calculus and Transformations 31

2 **Convex Optimization Problems** 35
   2.1 Convexity 36
   2.2 The C^1-Characterization of Convexity 41
      2.2.1 Multidimensional First Derivatives 41
      2.2.2 C^1-Characterization 44
   2.3 Solvability of Convex Optimization Problems 46
   2.4 Optimality Conditions for Unconstrained Convex Problems 47
   2.5 The C^2-Characterization of Convexity 51
      2.5.1 The Multidimensional Second Derivative 51
      2.5.2 C^2-Characterizations 53
   2.6 The Monotonicity Characterization of Convexity 57
   2.7 Optimality Conditions for Constrained Convex Problems 58
      2.7.1 Lagrange and Wolfe Duality 59
      2.7.2 The Karush-Kuhn-Tucker Conditions 72
      2.7.3 Complementarity 75
      2.7.4 Geometric Interpretation of the KKT Conditions 76
      2.7.5 Constraint Qualifications 78
   2.8 Algorithms 87
      2.8.1 Basic Idea of the Gradient Method 88
      2.8.2 Basic Idea of the Newton Method 89
      2.8.3 Basic Idea of Cutting Plane Methods 90
      2.8.4 Kelley's Cutting Plane Method 92
      2.8.5 The Frank-Wolfe Method 98
      2.8.6 Basic Idea of Primal-Dual Interior Point Methods 102

3 Nonconvex Optimization Problems ... 111
- 3.1 Examples and a Conceptual Algorithm ... 112
- 3.2 Convex Relaxation ... 115
- 3.3 Interval Arithmetic ... 120
  - 3.3.1 Motivation and Applications ... 121
  - 3.3.2 Basic Interval Operations ... 122
  - 3.3.3 Natural Interval Extension ... 126
  - 3.3.4 Dependency Effect ... 128
  - 3.3.5 Enclosure Property ... 129
  - 3.3.6 Taylor Models ... 131
  - 3.3.7 Further Notation ... 132
- 3.4 Convex Relaxation by the alphaBB Method ... 133
- 3.5 Uniformly Refined Tessellations ... 146
- 3.6 Branch-and-Bound for Box-Constrained Problems ... 154
- 3.7 Branch-and-Bound for Convexly Constrained Problems ... 162
- 3.8 Branch-and-Bound for Nonconvex Problems ... 163
- 3.9 Lipschitz Properties ... 168
  - 3.9.1 Properties of Lipschitz Continuous Functions ... 170
  - 3.9.2 Direct Application to Algorithm 3.5 ... 174
  - 3.9.3 A Variation of Algorithm 3.5 ... 176

References ... 179

Index ... 181

Introduction

Contents

1.1 Examples and Terminology 2
1.2 Solvability 12
    1.2.1 Definition of Solvability 12
    1.2.2 Types of Unsolvability 14
    1.2.3 The Weierstrass Theorem 18
    1.2.4 Unbounded Feasible Sets 20
    1.2.5 Nonclosed Feasible Sets 28
1.3 Rules of Calculus and Transformations 31

Finite-dimensional continuous optimization deals with the minimization or maximization of an objective function in a finite number of continuous decision variables. Important applications can be found not only in linear models (as in simple models for profit maximization in production programs or in transportation problems [30]), but also in various nonlinear models from natural, engineering, and economic sciences. These include geometric problems, mechanical problems, parameter-fitting problems, estimation problems, approximation problems, data classification, and sensitivity analysis. It is also used as a solution tool in noncooperative games [35], in robust optimization [35] or in the continuous relaxation of discrete and mixed-integer optimization problems [36].

This introductory chapter initially motivates in Sect. 1.1 the basic terminology and notation of optimization problems using various examples. Section 1.2 then extensively addresses the question, under what conditions optimization problems can be solved at all and what types of unsolvability can occur. Finally, Sect. 1.3 provides some rules of calculus and transformations for optimization problems, which play a role in this textbook.

© The Editor(s) (if applicable) and The Author(s), under exclusive license to Springer-Verlag GmbH, DE, part of Springer Nature 2024
O. Stein, *Basic Concepts of Global Optimization*, Mathematics Study Resources 5, https://doi.org/10.1007/978-3-662-66240-3_1

1.1 Examples and Terminology

In optimization, we compare different alternatives with respect to an objective criterion and search for a best among all considered alternatives. For example, one can ask who in a group of people carries the most coins. The alternatives are then the individual people, and the objective criterion is the number of coins a person carries. This objective criterion should be maximized.

Another example is a navigation system that determines a shortest road connection between two places. The alternatives are possible road connections, and the objective criterion is their respective length. It is to be minimized.

We refer to the set of considered alternatives as the *feasible set M*. For example, in the coin problem, one may only consider certain people, such as those in an age group or those of a gender. Similarly, in the navigation system, for example, only road connections that do not include toll roads or dirt roads may be considered.

Since an objective criterion assigns a number to each alternative x from M, it is a function f from the set M to the set of real numbers $\mathbb{R}$, in short

$$f : M \to \mathbb{R}, \quad x \mapsto f(x).$$

We call it the *objective function*.

The task of minimizing the function f over the set M is written as an optimization problem in the form

$$P : \quad \min \ f(x) \quad \text{s.t.} \quad x \in M.$$

The abbreviation s.t. stands for *subject to* (or *so that*) and indicates in the formulation of P that from here the description of the feasible set follows. A maximization problem would be written analogously in the form

$$P : \quad \max \ f(x) \quad \text{s.t.} \quad x \in M.$$

We will see, however, that it is sufficient to only deal with minimization problems.

An alternative $\bar{x} \in M$ is called *optimal* for P, if no alternative $x \in M$ has a better objective function value. In minimization problems, this means that the inequality $f(x) \geq f(\bar{x})$ is satisfied for all $x \in M$, and in maximization problems, this inequality is reversed. The associated *optimal value* of P is the number $v = f(\bar{x})$.

Often one denotes the optimal value of a minimization problem in the notation

$$v = \min_{x \in M} f(x).$$

Here, the 'min' in the above optimization problem P is to be understood as the *task* of minimizing f over M, while the 'min' in the optimal value v denotes a *number*.

1.1 Examples and Terminology

In the coin problem, the optimal value is the found maximal number of coins. It is uniquely determined. In contrast, several people can carry this number of coins. Therefore, the alternative $\bar{x}$ (here: the person), at which the optimal value is attained, is not necessarily unique.

This applies analogously to every optimization problem. For example, a navigation system not only provides the length of the shortest path between two places, but also a way to realize this length, namely a road connection as the optimal alternative. While the shortest possible length as the optimal value is unique, there may be several ways to realize this best length through a road connection.

In contrast to the coin problem and the navigation system, in which only finitely many (albeit possibly very many) feasible alternatives are compared, *continuous* optimization is characterized by the fact that M can contain a whole *continuum* of feasible alternatives, so in particular infinitely many alternatives. In finite-dimensional continuous optimization, we consider models where M is a subset of the finite-dimensional Euclidean space $\mathbb{R}^n$ (while, for example, in problems of optimal control M is a subset of a function space, so infinite-dimensional). Since the alternatives $x \in M \subseteq \mathbb{R}^n$ can be interpreted as points in the n-dimensional space, we will no longer call them alternatives, but points. In particular, the 'solution' of an optimization problem P will always consist in the specification of an *optimal point* $\bar{x} \in M$ and the associated *optimal value* $v = f(\bar{x})$.

Example 1.1.1 (Projection Onto a Set) For a set $M \subseteq \mathbb{R}^n$ and a point $z \in \mathbb{R}^n$, let the (Euclidean) distance from z to M be sought as well as a point $\bar{x}$ in M being closest to z (Fig. 1.1). The formulation as an optimization problem is

$$P : \quad \min_x \|x - z\|_2 \quad \text{s.t.} \quad x \in M.$$

The objective function here is $f(x, z) = \|x - z\|_2$, where the variables x and z play different roles: We want to *decide* on the choice of x, while z is exogenously given. The variable x is therefore called *decision variable* and z *parameter*. To clearly mark the decision variable, one can highlight it in the formulation of P as above by the notation $\min_x$. If one wants to clarify the ambient space of x at this opportunity, one can also write $\min_{x \in \mathbb{R}^n}$.

We call every optimal point $\bar{x}$ of P a *projection of z onto M*. When an optimal point $\bar{x}$ is found, the optimal value (here the distance from z to M) is simply calculated by substituting $\bar{x}$ into the objective function (here $v = \|\bar{x} - z\|_2$).

If we were interested in examining the dependence of the optimal point $\bar{x}$ and the optimal value v on the parameter z, we would explicitly note these as functions $\bar{x}(z)$ and $v(z)$. It would also be

Fig. 1.1 Projection onto a set in $\mathbb{R}^2$

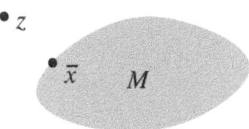

appropriate to refer to the optimization problem P more precisely as $P(z)$. However, since we do not want to investigate these parameter dependencies in the following and instead refer to [35], we do not explicitly note the corresponding functional dependencies.

Instead of the typical choice of the Euclidean norm in applications, one can use any other norm for distance measurement in the projection problem. This can have geometric reasons, for example, and generally leads to different results for the projection. To distinguish, when choosing the Euclidean norm, one also speaks of *orthogonal* projection.

In order to be able to handle an optimization problem P algorithmically, in addition to the objective function, the feasible set M must also be explicitly specified. In Example 1.1.1, the set M is only given *abstractly*. For the explicit description of a set M, various possibilities are conceivable, for two-dimensional sets for example a bitmap or the specification of its boundary as a polygonal chain. However, these approaches are not or only difficult to transfer to higher-dimensional problems.

Instead, the feasible sets of optimization problems are usually described with the help of equality and inequality constraints on x. For a simple case, namely the description of M by a single linear equality constraint, we modify Example 1.1.1.

Example 1.1.2 (Projection Onto a Hyperplane) In Example 1.1.1, let M be described by a single linear equality, i.e., as

$$M = \{x \in \mathbb{R}^n \mid a^\mathsf{T} x = b\}$$

with $a \neq 0$. Then M is referred to as *hyperplane*. In the case of $n = 3$, hyperplanes are planes and in the case of $n = 2$, they are lines (Fig. 1.2). The vector a always stands perpendicular to M and is therefore also called *normal vector*.

If there is an explicit description of M by constraints, it is sufficient to specify these constraints in the associated optimization problem P instead of M. To project a point $z \in \mathbb{R}^n$ onto the hyperplane M, we therefore have to solve the problem

$$P : \min_{x \in \mathbb{R}^n} \|x - z\|_2 \quad \text{s.t.} \quad a^\mathsf{T} x = b$$

(i.e., specify the optimal value and an optimal point).

This is not always explicitly possible for general optimization problems, even when M is specified by constraints. In the present example, however, we will

Fig. 1.2 Projection onto a line

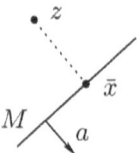

1.1 Examples and Terminology

Fig. 1.3 Nonunique optimal point

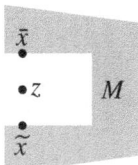

Fig. 1.4 Local and global minimal point

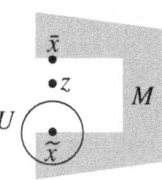

succeed with the techniques from Chap. 2: The unique optimal point is calculated as

$$\bar{x} = z - \frac{a^\mathsf{T} z - b}{a^\mathsf{T} a} \cdot a,$$

and the associated optimal value is

$$v = \left\| \left(z - \frac{a^\mathsf{T} z - b}{a^\mathsf{T} a} \cdot a \right) - z \right\|_2 = \frac{|a^\mathsf{T} z - b|}{\|a\|_2}$$

(Example 2.7.11). Specifically for the case $z = 0$, $\bar{x} = (b/a^\mathsf{T} a)a$ is referred to as the *norm-minimal solution* of the equation $a^\mathsf{T} x = b$. This concept can easily be extended to underdetermined *systems* of equations.

An example of a nonunique optimal point in the projection problem is given in Fig. 1.3. Here, both $\bar{x}$ and $\tilde{x}$ have minimal distance from z among all points of the set M, so there are two different projections of z onto M.

If one moves the point z in the direction of $\bar{x}$ in Fig. 1.3, $\bar{x}$ becomes the unique minimal point. The point $\tilde{x}$ retains the property that there are no points in M in a sufficiently small neighborhood U around $\tilde{x}$ that are closer to z than $\tilde{x}$ (Fig. 1.4). These two situations are distinguished by calling $\bar{x}$ a *global* minimal point and $\tilde{x}$ a *local* minimal point.

This important distinction is formally established by the following definition (Fig. 1.5).

Definition 1.1.3 (Minimal Points and Minimal Values) Let a feasible set $M \subseteq \mathbb{R}^n$ and an objective function $f : M \to \mathbb{R}$ be given.

(continued)

(a) $\bar{x} \in M$ is called a *local minimal point* of f on M, if there exists a neighborhood U of $\bar{x}$ with

$$\forall x \in U \cap M : \quad f(x) \geq f(\bar{x}).$$

(b) $\bar{x} \in M$ is called a *global minimal point* of f on M, if in part a $U = \mathbb{R}^n$ can be chosen.
(c) A local or global minimal point is called *strict*, if in part a or part b for $x \neq \bar{x}$ even the strict inequality $>$ holds.
(d) For each global minimal point $\bar{x}$, $f(\bar{x})$ ($= v = \min_{x \in M} f(x)$) is called *global minimal value*, and for each local minimal point $\bar{x}$, $f(\bar{x})$ is called *local minimal value*.

Regarding the definition of minimal points and values, we note the following:

- For the requirement $f(x) \geq f(\bar{x})$ to make sense, the image space of f must be ordered. For example, the minimization of $f : \mathbb{R}^n \to \mathbb{R}^2$ is initially not meaningful. However, the field of *multicriteria optimization* deals with how such problems can still be handled (for a brief introduction see, for example, [30]).
- Every global minimal point is also a local minimal point.
- Strict global minimal points are unique, and strict local minimal points are locally unique.
- Local and global *maximal points* are defined analogously. Since the maximal points of f are exactly the minimal points of $-f$, it is sufficient to consider minimization problems. Attention: Because of $\max f(x) = -\min(-f(x))$ this

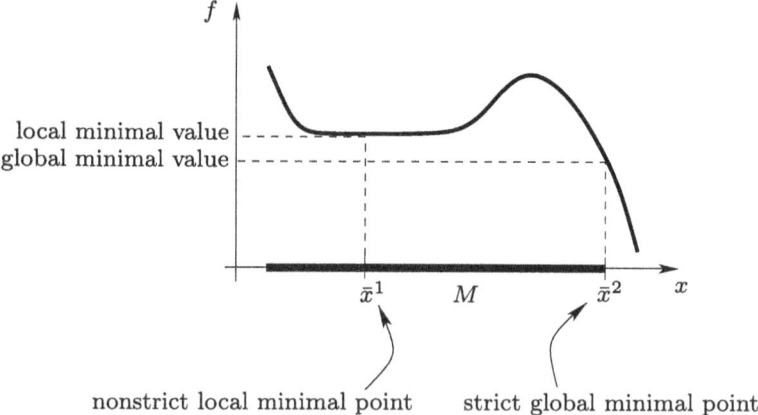

Fig. 1.5 Local and global minimality

Fig. 1.6 Maximization of f through minimization of $-f$

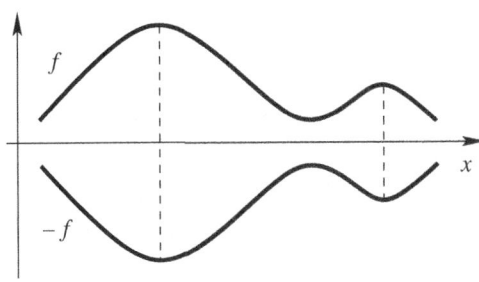

Fig. 1.7 Point cloud

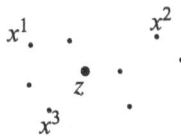

construction changes the sign of the optimal *value*. This is illustrated in Fig. 1.6 and proven in Exercise 1.1.4 as well as more generally in Exercise 1.3.1.
- Due to the similar notation, there is a risk of confusion between the minimal *value* $\min_{x \in M} f(x)$ and the underlying minimization *task* P (see the discussion at the beginning of this section).

Exercise 1.1.4 Let a feasible set $M \subseteq \mathbb{R}^n$ and an objective function $f : M \to \mathbb{R}$ be given. Show:

(a) The global maximal points of f on M are exactly the global minimal points of $-f$ on M.
(b) Provided f has global maximal points, the global maximal value satisfies

$$\max_{x \in M} f(x) = -\min_{x \in M} (-f(x)).$$

Example 1.1.5 (Center of a Point Cloud) Given points $x^1, x^2, \ldots, x^m \in \mathbb{R}^n$ (i.e. $x^i = (x_1^i, \ldots, x_n^i)^T$, $i = 1, \ldots, m$), we look for a point $z \in \mathbb{R}^n$ 'in the center of $x^1, x^2, \ldots, x^m$' (Fig. 1.7).

First, we need to clarify how to define 'center'. An obvious requirement is that for a center point z, the occurring distances $\|z - x^1\|_2, \|z - x^2\|_2, \ldots, \|z - x^m\|_2$ should all simultaneously be close to zero. However, if we were to minimize each of these expressions separately, we would end up with a multicriteria problem with m objective functions. Instead, we take advantage of the definiteness and continuity of norms, according to which the aforementioned individual expressions are all simultaneously close to zero if the norm of their vector

$$\left\| \begin{pmatrix} \|z - x^1\|_2 \\ \vdots \\ \|z - x^m\|_2 \end{pmatrix} \right\|_2$$

is as small as possible (i.e., a single function). This leads to the optimization problem

$$P : \min_{z \in \mathbb{R}^n} \left\| \begin{pmatrix} \|z - x^1\|_2 \\ \vdots \\ \|z - x^m\|_2 \end{pmatrix} \right\|_2 .$$

Here z is the decision variable, and the points $x^1, x^2, \ldots, x^m$ could be considered as parameters if necessary (which is not the case here, so we do not explicitly note this dependency). P has no constraints. The unique optimal point turns out to be the arithmetic mean of the data points (for the justification see Sec. 2.4):

$$\bar{z} = \frac{1}{m} \sum_{i=1}^{m} x^i.$$

In some applications, it may also make sense to choose z as one of the points $x^1, x^2, \ldots, x^m$ in order to consider z as a 'typical' point. The associated optimization problem then has the constraint $z \in \{x^1, x^2, \ldots, x^m\}$ and thus falls into the class of discrete optimization problems, which we do not cover in this textbook (instead see, e.g., [36]).

A problem with the approach in Example 1.1.5 is that outliers are increasingly neglected as the number of points m increases. This can be desirable (e.g., if measurement errors in the points must be assumed) or undesirable (e.g., if the points correspond to geographical locations and remote locations should be treated in a fair manner; Fig. 1.8).

This can be prevented by choosing a different (outer) norm, for example the maximum norm

$$\|a\|_\infty = \max_{i=1,\ldots,n} |a_i|$$

Fig. 1.8 Outlier x^1 and various centers

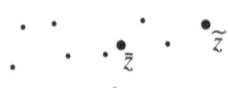

1.1 Examples and Terminology

for $a \in \mathbb{R}^n$ (also called ℓ_∞-norm or Chebyshev norm). By choosing this norm, the center $\tilde{z}$ can be considered as the center of the smallest possible ball that contains all x^i. To see this, we introduce the epigraph of a function, which will also serve us well in other contexts. In its and some later definitions we will not denote the domain of f by M, but by X, since it does not necessarily have to be the feasible set of an optimization problem.

Definition 1.1.6 (Epigraph) For $X \subseteq \mathbb{R}^n$ and $f : X \to \mathbb{R}$ the set

$$\mathrm{epi}(f, X) = \{(x, \alpha) \in X \times \mathbb{R} |\ f(x) \le \alpha\}$$

is called the *epigraph* of f on X.

The epigraph consists of the graph of f on X and all points above it (Fig. 1.9). It is easy to see (Exercise 1.3.7) that one can find a minimal point of f on a feasible set M by looking for a point (x, α) with minimal α-component in the epigraph of f on M. The problem

$$P : \quad \min_{x} f(x) \quad \text{s.t.} \quad x \in M$$

is in this sense equivalent to

$$P_{\mathrm{epi}} : \quad \min_{(x,\alpha)} \alpha \quad \text{s.t.} \quad f(x) \le \alpha, \quad x \in M.$$

This *epigraph reformulation* of P can have advantages, for example, when the function f is the maximum of other functions (Example 1.1.7) or when one is forced to use a linear objective function (e.g., Sect. 2.8.4 and Exercise 3.2.6).

Example 1.1.7 (Center of a Point Cloud—Sequel 1) In the case of the center of a point cloud from Example 1.1.5, with the maximum norm as the outer norm, we

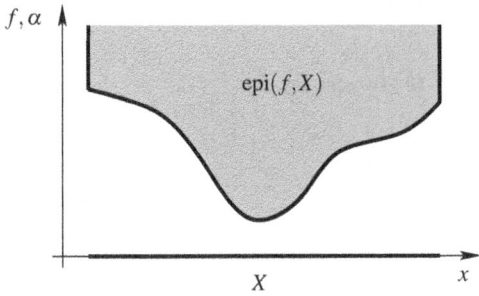

Fig. 1.9 Epigraph of f on X

have

$$P: \min_{z \in \mathbb{R}^n} \left\| \begin{pmatrix} \|z - x^1\|_2 \\ \vdots \\ \|z - x^m\|_2 \end{pmatrix} \right\|_\infty,$$

so $M = \mathbb{R}^n$ and

$$f(z) = \max_{i=1,\ldots,m} \|z - x^i\|_2.$$

As the epigraph reformulation we obtain the equivalent problem

$$P_{\text{epi}}: \min_{(z,\alpha)} \alpha \quad \text{s.t.} \quad f(z) \le \alpha,$$

where the constraint of P_{epi} more explicitly reads

$$\max_{i=1,\ldots,m} \|z - x^i\|_2 \le \alpha.$$

Since the maximum of m numbers lies below some upper bound α if and only if all m numbers lie below α, this constraint can be equivalently reformulated to

$$\|z - x^i\|_2 \le \alpha, \quad i = 1, \ldots, m,$$

and we obtain the equivalence of P to

$$P_{\text{epi}}: \min_{(z,\alpha)} \alpha \quad \text{s.t.} \quad \|z - x^i\|_2 \le \alpha, \quad i = 1, \ldots, m.$$

If one considers α as a radius, one thus tries to find the ball with center $\tilde{z}$ and minimal radius $\tilde{\alpha}$ that contains all points $x^1, \ldots, x^m$.

Example 1.1.8 (Cluster Analysis) If in Example 1.1.5 it is expected that the points $x^1, x^2, \ldots, x^m$ accumulate at q positions, one can try to determine 'q centers simultaneously' (Fig. 1.10).

Fig. 1.10 Two clusters

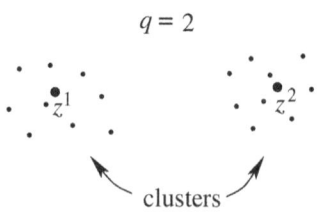

1.1 Examples and Terminology

If the centers are called $z^1, \ldots, z^q$, then for each z^ℓ there is a cluster C^ℓ, $\ell = 1, \ldots, q$, which consists of the nearest points. A crucial question is to which cluster a point x^i belongs.

Let the index of this cluster be $\ell(i)$. Then $x^i \in C^{\ell(i)}$ holds if and only if x^i has a smaller distance from $z^{\ell(i)}$ than from all other z^ℓ, i.e. when

$$\| z^{\ell(i)} - x^i \| = \min_{\ell=1,\ldots,q} \| z^\ell - x^i \|$$

applies.

The distances occurring in the problem of cluster analysis are therefore $\| z^{\ell(i)} - x^i \|$, $i = 1, \ldots, m$, and the associated optimization problem is

$$P: \min_{z^1,\ldots,z^q \in \mathbb{R}^n} f(z^1, \ldots, z^q)$$

with the objective function

$$f(z^1,\ldots,z^q) = \left\| \begin{pmatrix} \| z^{\ell(1)} - x^1 \| \\ \vdots \\ \| z^{\ell(m)} - x^m \| \end{pmatrix} \right\| = \left\| \begin{pmatrix} \min_{\ell=1,\ldots,q} \| z^\ell - x^1 \| \\ \vdots \\ \min_{\ell=1,\ldots,q} \| z^\ell - x^m \| \end{pmatrix} \right\|.$$

As the 'outer' norm, often the ℓ_1-norm is often used ($\|a\|_1 = \sum_{i=1}^n |a_i|$ for $a \in \mathbb{R}^n$; aka Manhattan norm), so that P takes the form

$$P_1: \min_{z^1,\ldots,z^q \in \mathbb{R}^n} \sum_{i=1}^m \min_{\ell=1,\ldots,q} \| z^\ell - x^i \|,$$

or also the ℓ_∞-norm with

$$P_\infty: \min_{z^1,\ldots,z^q \in \mathbb{R}^n} \max_{i=1,\ldots,m} \min_{\ell=1,\ldots,q} \| z^\ell - x^i \|.$$

In any case, the vector of decision variables in the problem of cluster analysis has the dimension $n \cdot q$, so that in practically relevant applications, high-dimensional problems often occur.

The algorithmic solution of such problems to global optimality is considered difficult, not only because of the possible high-dimensionality, but especially due to the 'minimum structure' in the objective function, which leads to nonconvexity (Chap. 2).

1.2 Solvability

Whether an optimization problem has optimal points at all, is not always obvious, and in many solution methods it must be checked or guaranteed in advance by the user. This is the subject of the present section. After the definition of the concept of solvability in Sect. 1.2.1, Sect. 1.2.2 first clarifies which types of unsolvability can occur. With the Weierstrass theorem, Sect. 1.2.3 formulates the central sufficient condition for the solvability of optimization problems. Since its assumption of a bounded and closed feasible set is violated in many applications, Sects. 1.2.4 and 1.2.5 provide modifications of these conditions for unbounded or nonclosed feasible sets.

We follow a construction common in mathematical texts for negations and use, for example, the artificial term 'nonclosed' instead of 'not closed', so it is clear what the 'not' refers to.

1.2.1 Definition of Solvability

Without any assumptions on the set $M \subseteq \mathbb{R}^n$ and the function $f : M \to \mathbb{R}$, to each minimization problem

$$P: \quad \min f(x) \quad \text{s.t.} \quad x \in M$$

a 'generalized minimal value' can be assigned, namely the *infimum* of f on M. To introduce it formally, we call $\alpha \in \mathbb{R}$ a *lower bound* for f on M, if

$$\forall\, x \in M: \quad \alpha \le f(x)$$

holds. The infimum of f on M is the *greatest* lower bound of f on M, so $v = \inf_{x \in M} f(x)$, if

- $v \le f(x)$ for all $x \in M$ (i.e., v is itself a lower bound of f on M) and
- $\alpha \le v$ for all lower bounds α of f on M.

Similarly, the *supremum* $\sup_{x \in M} f(x)$ of f on M is defined as the *smallest upper bound*.

Example 1.2.1 It holds $\inf_{x \in \mathbb{R}} (x - 5)^2 = 0$ and $\inf_{x \in \mathbb{R}} e^x = 0$.

If f is not bounded from below on M, one formally sets

$$\inf_{x \in M} f(x) = -\infty,$$

1.2 Solvability

and for the infimum over the empty set, one formally defines

$$\inf_{x \in \emptyset} f(x) = +\infty$$

(where the properties of f then play no role). We will see after Theorem 1.2.9 why these formal settings make sense.

Example 1.2.2 It holds $\inf_{x \in \mathbb{R}} (x - 5) = -\infty$ and $\inf_{x \in \emptyset}(x - 5) = +\infty$.

The 'generalized minimal value' $\inf_{x \in M} f(x)$ of P is always an element of the *extended real numbers* $\overline{\mathbb{R}} := \mathbb{R} \cup \{\pm\infty\}$. In analysis it is shown (e.g. [18]), that the so defined infimum exists without any assumptions on f and M and is uniquely determined.

> **Definition 1.2.3 (Solvability)** A minimization problem P is called *solvable*, if there exists some $\bar{x} \in M$ with $\inf_{x \in M} f(x) = f(\bar{x})$.

Solvability of P thus means that the infimum of f on M can be realized as the objective function value of some feasible point, i.e., the infimum is *attained*. To indicate that the infimum is attained we write $\min_{x \in M} f(x)$ instead of $\inf_{x \in M} f(x)$.

Example 1.2.4 It holds $0 = \min_{x \in \mathbb{R}} (x - 5)^2 = (\bar{x} - 5)^2$ with $\bar{x} = 5$, but there is no $\bar{x} \in \mathbb{R}$ with $0 = \inf_{x \in \mathbb{R}} e^x = e^{\bar{x}}$.

The following theorem states (unsurprisingly) that for solvability one can equivalently require the existence of a global minimal point.

> **Theorem 1.2.5** *A minimization problem P is solvable if and only if it has a global minimal point.*

Proof First, let P be solvable. Then there exists some $\bar{x} \in M$ with $\min_{x \in M} f(x) = f(\bar{x})$. As an infimum, $f(\bar{x})$ is a lower bound for f on M, so it holds $f(\bar{x}) \leq f(x)$ for all $x \in M$. According to Definition 1.1.3, $\bar{x}$ is therefore a global minimal point of f on M.

On the other hand, let $\bar{x}$ be a global minimal point of f on M. Then $\bar{x} \in M$ holds, and $f(\bar{x})$ is a lower bound for f on M. If there was a greater lower bound α for f on M, we would have

$$\forall x \in M : \quad f(\bar{x}) < \alpha \leq f(x),$$

which leads to a contradiction for $x = \bar{x}$. Therefore, $f(\bar{x})$ is the greatest lower bound for f on M, and $\inf_{x \in M} f(x) = f(\bar{x})$ holds. □

In analysis (e.g. [18]), the following and subsequently employed result is proven for the characterization of infima: The infimum of a nonempty set of real numbers is exactly that one of its lower bounds which can be approximated arbitrarily well by elements from the set. For the infima of functions on sets considered here, this means that, for $M \neq \emptyset$, $v = \inf_{x \in M} f(x)$ holds if and only if $v \leq f(x)$ is true for all $x \in M$ and a sequence $(x^k) \subseteq M$ with $v = \lim_k f(x^k)$ exists. Here and in the following, we write briefly (x^k) for a sequence $(x^k)_{k \in \mathbb{N}}$ and $\lim_k$ for $\lim_{k \to \infty}$.

Exercise 1.2.6 Show for every nonempty set $M \subseteq \mathbb{R}^n$ and every function $f : M \to \mathbb{R}$ the identity

$$\{\alpha \in \mathbb{R}|\ \forall x \in M : \alpha \leq f(x)\} = \{\alpha \in \mathbb{R}|\ \alpha \leq \inf_{x \in M} f(x)\}.$$

1.2.2 Types of Unsolvability

Before we turn to sufficient criteria for the *solvability* of P, we first address the question of which *types of unsolvability* are possible. This is interesting, for example, for algorithms that not only solve P, but also provide a corresponding output message in the case of unsolvability (such as the simplex algorithm of linear optimization [30]).

For this, we consider the *parallel projection* of the epigraph

$$\text{epi}(f, M) = \{(x, \alpha) \in M \times \mathbb{R}|\ f(x) \leq \alpha\}$$

of f on M onto the 'α-axis'. It is related to the orthogonal projection of a point z onto a set M from Example 1.1.1 as follows.

Exercise 1.2.7 We denote with $\text{pr}(z, M)$ the set of orthogonal projections of a point $z \in \mathbb{R}^n$ onto $M \subseteq \mathbb{R}^n$ and with

$$\text{pr}(Z, M) = \{\text{pr}(z, M)|\ z \in Z\}$$

the orthogonal projection of a set $Z \subseteq \mathbb{R}^n$ onto M.

We now consider the special case $M = \mathbb{R}^k \times \{0_\ell\}$ with $0_\ell \in \mathbb{R}^\ell$ and $k + \ell = n$. For this, we split the vector x into $x = (a, b)$ with $a \in \mathbb{R}^k$ and $b \in \mathbb{R}^\ell$. Show for every set $Z \subseteq \mathbb{R}^n = \mathbb{R}^k \times \mathbb{R}^\ell$ the identity

$$\text{pr}(Z, \mathbb{R}^k \times \{0_\ell\}) = \{(a, 0_\ell) \in \mathbb{R}^k \times \{0_\ell\}|\ \exists b \in \mathbb{R}^\ell : (a, b) \in Z\}.$$

1.2 Solvability

The omission of the set $\{0_\ell\}$ in the identity for $\mathrm{pr}(Z, \mathbb{R}^k \times \{0_\ell\})$ from Exercise 1.2.7 motivates the following definition.

Definition 1.2.8 (Parallel Projection) For $Z \subseteq \mathbb{R}^n = \mathbb{R}^k \times \mathbb{R}^\ell$, we call

$$\mathrm{pr}_a(Z) = \{a \in \mathbb{R}^k | \exists b \in \mathbb{R}^\ell : (a,b) \in Z\}$$

the *parallel projection* of Z onto the set $\mathbb{R}^k$, somewhat loosely referred to as the 'a-space'.

The parallel projection of $\mathrm{epi}(f, M)$ onto the 'α-space' $\mathbb{R}$ is therefore

$$\mathrm{pr}_\alpha \mathrm{epi}(f, M) = \{\alpha \in \mathbb{R} | \exists x \in M : f(x) \leq \alpha\}.$$

Figures 1.11 and 1.12 show the sets $\mathrm{pr}_\alpha \mathrm{epi}(f, M)$ for the two optimization problems from Example 1.2.1. In Example 1.2.2 we have $\mathrm{pr}_\alpha \mathrm{epi}(x-5, \mathbb{R}) = \mathbb{R}$ and $\mathrm{pr}_\alpha \mathrm{epi}(x-5, \emptyset) = \emptyset$.

The following theorem states that also in general $\mathrm{pr}_\alpha \mathrm{epi}(f, M)$ can possess only one of exactly four shapes, exactly one of which corresponds to the solvability of P.

Fig. 1.11 The set $\mathrm{pr}_\alpha \mathrm{epi}(f, \mathbb{R})$ for $f(x) = (x-5)^2$

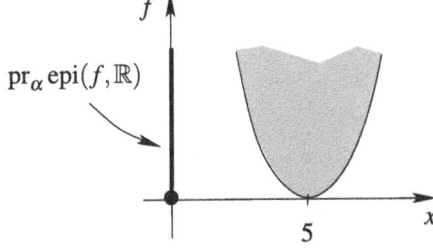

Fig. 1.12 The set $\mathrm{pr}_\alpha \mathrm{epi}(f, \mathbb{R})$ for $f(x) = e^x$

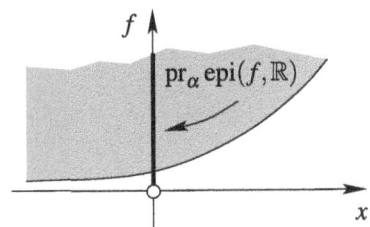

Theorem 1.2.9 *For each optimization problem P, the set $\mathrm{pr}_\alpha \mathrm{epi}(f, M)$ possesses exactly one of the following four shapes:*

(a) $\mathrm{pr}_\alpha \mathrm{epi}(f, M) = \emptyset$
(b) $\mathrm{pr}_\alpha \mathrm{epi}(f, M) = \mathbb{R}$
(c) $\mathrm{pr}_\alpha \mathrm{epi}(f, M) = (v, +\infty)$ with $v \in \mathbb{R}$
(d) $\mathrm{pr}_\alpha \mathrm{epi}(f, M) = [v, +\infty)$ with $v \in \mathbb{R}$

Furthermore, case a is equivalent to $M = \emptyset$, case b to the unboundedness of f on M from below, case c to the fact that the finite infimum $v = \inf_{x \in M} f(x)$ is not attained, and case d to the solvability of P with $v = \min_{x \in M} f(x)$.

Proof For abbreviation, we set $V := \mathrm{pr}_\alpha \mathrm{epi}(f, M)$ and first distinguish whether in P the feasible set M is empty or not. For $M = \emptyset$ it follows $V = \emptyset$, so it is shown that case a can occur. To see that case a *only* occurs for $M = \emptyset$, choose in the case $M \neq \emptyset$ some $\bar{x} \in M$ and set $\bar{\alpha} = f(\bar{x})$. Then $\bar{\alpha} \in V$ and thus $V \neq \emptyset$ hold.

In the following let $M \neq \emptyset$ and thus $V \neq \emptyset$. Choose an arbitrary $\alpha \in V$. Then all $\tilde{\alpha} \geq \alpha$ lie also in V, because there is an $x \in M$ with $f(x) \leq \alpha \leq \tilde{\alpha}$. From this follows

$$V \supseteq \bigcup_{\alpha \in V} [\alpha, +\infty).$$

Now let f be unbounded from below on M. Then there is a sequence $(x^k) \subseteq M$ with $f(x^k) \leq -k$ for all $k \in \mathbb{N}$. In particular, we have $-k \in V$ for all $k \in \mathbb{N}$ and thus

$$V \supseteq \bigcup_{\alpha \in V} [\alpha, +\infty) \supseteq \bigcup_{k \in \mathbb{N}} [-k, +\infty) = \mathbb{R},$$

so case b. To see that this case *only* occurs for f unbounded from below on M, choose in the case of a function f bounded from below on M a lower bound $\bar{\alpha} \in \mathbb{R}$ with $\bar{\alpha} \leq f(x)$ for all $x \in M$. Now choose for some arbitrary $\alpha \in V$ an arbitrary $x \in M$ with $f(x) \leq \alpha$. Then $\bar{\alpha} \leq f(x) \leq \alpha$ and therefore $V \subseteq [\bar{\alpha}, +\infty)$ are true. Due to $\bar{\alpha} \in \mathbb{R}$ this excludes case b.

In the following let f be bounded from below on $M \neq \emptyset$ by $\bar{\alpha} \in \mathbb{R}$. Then the greatest lower bound of f on M is also a real number, so $v := \inf_{x \in M} f(x) \in \mathbb{R}$ exists and there exists a sequence $(x^k) \subseteq M$ with $\lim_k f(x^k) = v$, where the sequence $(f(x^k))$ can be assumed to be strictly monotonically decreasing without loss of generality. From $(f(x^k)) \subseteq V$ follows

$$[v, +\infty) \supseteq V \supseteq \bigcup_{\alpha \in V} [\alpha, +\infty) \supseteq \bigcup_{k \in \mathbb{N}} [f(x^k), +\infty) = (v, +\infty).$$

For the shape of the set V this results in case c or d. If v is attained as the infimum, there is some $\bar{x} \in M$ with $f(\bar{x}) = v$, so that $v \in V$ and thus $V = [v, +\infty)$ applies, so case d. If v is not attained as the infimum, it results analogously $v \notin V$ and thus case c. □

According to Theorem 1.2.9 there are exactly three reasons for the unsolvability of P, namely *inconsistency* of P (more precisely: of M or, rather, the description of M) in case a, the *unboundedness* of P (more precisely: the unboundedness from

1.2 Solvability

below of f on M) in case b and the fact that a *finite infimum* (i.e. $v \notin \{\pm\infty\}$) *is not attained* in case c.

Inconsistent optimization problems and feasible sets are also called *infeasible*, but due to its ambiguity we do not adopt this terminology.

In particular, P is unsolvable if and only if $\mathrm{pr}_\alpha \mathrm{epi}(f, M)$ is an open set. Since the sets in case a and b are also closed, the alternative c is eliminated for all classes of optimization problems in which $\mathrm{pr}_\alpha \mathrm{epi}(f, M)$ is always a closed set. This is the case, for example, in linear optimization (because then $\mathrm{epi}(f, M)$ is a convex polyhedron, and parallel projections of convex polyhedra are again convex polyhedra and, thus, in particular closed; e.g. [40]). In fact, in case of unsolvability the simplex algorithm of linear optimization either reports inconsistency or unboundedness of P.

If one makes use of the fact that the infimum $v = \inf_{x \in M} f(x)$ is an element of the extended real numbers $\overline{\mathbb{R}}$, the classification in Theorem 1.2.9 becomes even more transparent, and also the motivation for the formal settings of the infimum in the cases of inconsistency and unboundedness becomes clear. In the four cases from Theorem 1.2.9 it holds:

(a) $\mathrm{pr}_\alpha \mathrm{epi}(f, M) = (v, +\infty)$ with $v = +\infty$
(b) $\mathrm{pr}_\alpha \mathrm{epi}(f, M) = (v, +\infty)$ with $v = -\infty$
(c) $\mathrm{pr}_\alpha \mathrm{epi}(f, M) = (v, +\infty)$ with $v \in \mathbb{R}$
(d) $\mathrm{pr}_\alpha \mathrm{epi}(f, M) = [v, +\infty)$ with $v \in \mathbb{R}$

According to Theorem 1.2.9, v in case c and d corresponds to the infimum of f on M. The above representation explains why one analogously sets in case a $\inf_{x \in \emptyset} f(x) = +\infty$ and in case b $\inf_{x \in M} f(x) = -\infty$. The *calculation* with these extended real 'numbers' is not straightforward, but some subsequent inequalities involving infima $v \in \{\pm\infty\}$ will at least motivate facts that we will then prove properly.

Whether unsolvability is actually problematic depends on the application, as the following example shows.

Example 1.2.10 (Distance of a Point from a Set) For a set $M \subseteq \mathbb{R}^n$ and a point $z \in \mathbb{R}^n$, the term

$$\mathrm{dist}(z, M) := \inf_{x \in M} \|x - z\|_2$$

is called the *distance* of z from M. As seen in Example 1.1.1, every point $\bar{x} \in M$, whose objective function value $\|\bar{x} - z\|_2$ realizes the distance, is called a projection of z onto M. However, the distance is also meaningfully defined when such a projection does not exist because the underlying optimization problem is unsolvable due to nonattainment of a finite infimum. This is illustrated in Fig. 1.13 for a nonclosed set M, i.e., there is a sequence $(x^k) \subseteq M$, which has a limit point $x^* \notin M$. An optimal point does not exist here because every feasible point $x \in M$ can be improved.

Fig. 1.13 Distance to a nonclosed set

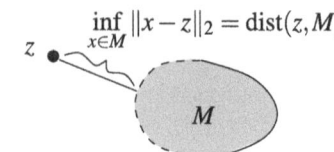

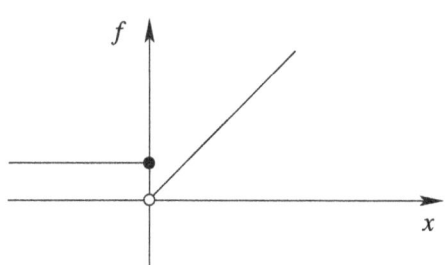

Fig. 1.14 Unsolvability due to discontinuity of f

Due to the nonnegativity of the norm, unsolvability due to unboundedness is not possible for this problem. Furthermore, $\text{dist}(z, \emptyset) = +\infty$ holds for every $z \in \mathbb{R}^n$.

1.2.3 The Weierstrass Theorem

Next we turn to sufficient conditions for the solvability of P. A minimal requirement for this is obviously the consistency of M, i.e. $M \neq \emptyset$. In Example 1.2.10 we also saw that missing closedness of M can lead to unsolvability.

The unsolvability of the minimization of e^x on $\mathbb{R}$ is due to an asymptotic effect for $x \to -\infty$, which can be excluded if one additionally requires the boundedness of M, i.e., the existence of some $R > 0$ with $M \subseteq \{x \in \mathbb{R}^n | \, \|x\| \leq R\}$ (i.e., M lies within a sufficiently large ball around the origin; the choice of the norm is irrelevant). A set $M \subseteq \mathbb{R}^n$ that is simultaneously closed and bounded is called *compact*.

Finally, unsolvability can also be due to discontinuity of the objective function f. For example,

$$f(x) = \begin{cases} 1, & x \leq 0 \\ x, & x > 0 \end{cases}$$

has no global minimal point on $\mathbb{R}$, because again there is an improvement possibility for every solution candidate (Fig. 1.14). This situation is excluded by the requirement of continuity of f.

The following central theorem on the existence of minimal and maximal points shows that the listed requirements on f and M are indeed sufficient to guarantee the solvability of P.

1.2 Solvability

Theorem 1.2.11 (Weierstrass Theorem) *Let the set $M \subseteq \mathbb{R}^n$ be nonempty and compact, and let the function $f : M \to \mathbb{R}$ be continuous. Then f possesses (at least) one global minimal point and one global maximal point on M.*

Proof Let $v = \inf_{x \in M} f(x)$. In view of $M \neq \emptyset$ we have $v < +\infty$. It remains to show the existence of some $\bar{x}$ in M with $v = f(\bar{x})$. Since v is the infimum, there exists a sequence $(x^k) \subseteq M$ with $\lim_k f(x^k) = v$. In analysis it is proven (in the Bolzano-Weierstrass theorem; e.g. [17]), that every sequence (x^k) lying in a compact set M has a convergent subsequence in M. To avoid having to use a tedious subsequence notation, we choose our sequence (x^k) directly as such a convergent sequence, so there exists an $x^\star \in M$ with $\lim_k x^k = x^\star$. The continuity of f on M yields

$$f(x^\star) = f\left(\lim_k x^k\right) = \lim_k f(x^k) = v,$$

so we can choose $\bar{x} := x^\star$. The proof for the existence of a global maximal point proceeds analogously. □

Example 1.2.12 (Projection Onto a Set—Sequel 1) According to Theorem 1.2.11, the projection problem from Example 1.1.1 is solvable for every nonempty and compact set $M \subseteq \mathbb{R}^n$, because $f(x, z) = \|x - z\|_2$ is a continuous function in x for every $z \in \mathbb{R}^n$.

While the Weierstrass theorem provides the central *sufficient* conditions for the solvability of an optimization problem, it will be crucial for the following that apart from the consistency of M, none of these conditions are also *necessary* for solvability. That the conditions are stronger than necessary is already evident from the fact that the Weierstrass theorem also guarantees the existence of a global *maximal* point, which we were not interested in.

For example, the existence of minimal points (but not maximal points) is also guaranteed for certain discontinuous functions f. A discussion of this *lower semi-continuity* can be found, for example, in [35].

In the following, we will instead examine how the boundedness and closedness of M can be weakened, as these assumptions are often violated in application problems. Especially for problems without constraints, i.e., *unconstrained problems*, $M = \mathbb{R}^n$ applies (e.g., for computing the center of a point cloud, Example 1.1.5, and in cluster analysis, Example 1.1.8). Although M is then nonempty and closed, it is *not* bounded. Therefore, Theorem 1.2.11 is not applicable to unconstrained problems.

1.2.4 Unbounded Feasible Sets

To make the Weierstrass theorem applicable to problems with an unbounded set M, we use a 'trick' and consider lower level sets of f.

> **Definition 1.2.13 (Lower Level Set)** For $X \subseteq \mathbb{R}^n$, $f : X \to \mathbb{R}$ and $\alpha \in \overline{\mathbb{R}}$,
> $$\mathrm{lev}_{\leq}^{\alpha}(f, X) = \{x \in X| \ f(x) \leq \alpha\}$$
> is called the *lower level set of f on X at level α*. In the case $X = \mathbb{R}^n$, we also write briefly
> $$f_{\leq}^{\alpha} := \mathrm{lev}_{\leq}^{\alpha}(f, \mathbb{R}^n) \quad (= \{x \in \mathbb{R}^n| \ f(x) \leq \alpha\}).$$

For every real-valued function f, it holds $\mathrm{lev}_{\leq}^{+\infty}(f, X) = X$ and $\mathrm{lev}_{\leq}^{-\infty}(f, X) = \emptyset$. The set $\mathrm{lev}_{\leq}^{\alpha}(f, X)$ must not be confused with the epigraph of f on X,

$$\mathrm{epi}(f, X) = \{(x, \alpha) \in X \times \mathbb{R}| \ f(x) \leq \alpha\}.$$

Example 1.2.14 For $f(x) = x^2$, it holds $f_{\leq}^1 = [-1, 1]$, $f_{\leq}^0 = \{0\}$ and $f_{\leq}^{-1} = \emptyset$ (Fig. 1.15), and for $f(x) = x_1^2 + x_2^2$, it holds $f_{\leq}^1 = \{x \in \mathbb{R}^2| \ x_1^2 + x_2^2 \leq 1\}$ (the unit disk), $f_{\leq}^0 = \{0\}$ and $f_{\leq}^{-1} = \emptyset$ (Fig. 1.16).

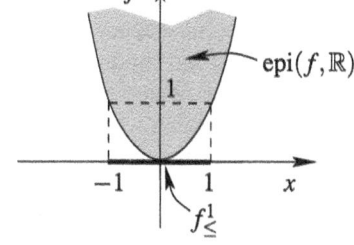

Fig. 1.15 Lower level set $f_{\leq}^1$ and epigraph of $f(x) = x^2$ on $\mathbb{R}$

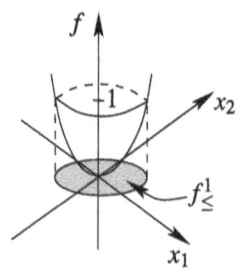

Fig. 1.16 Lower level set $f_{\leq}^1$ of $f(x) = x_1^2 + x_2^2$ on $\mathbb{R}^2$

1.2 Solvability

In analysis (and in [37]), it is shown that for continuous f, the sets $f_\leq^\alpha$ are closed for all $\alpha \in \mathbb{R}$.

In the following results we denote the set of global minimal points of P by

$$S = \{\bar{x} \in M \mid \forall x \in M : f(x) \geq f(\bar{x})\}.$$

The solvability of P is then equivalent to $S \neq \emptyset$; however, we will be able to show further properties of the set S.

Lemma 1.2.15 *Let $v = \inf_{x \in M} f(x)$. Then $S = \mathrm{lev}_\leq^v(f, M)$ holds.*

Proof The assertion follows from the chain of equivalences

$$\begin{array}{rcl}
\bar{x} \in S & \Leftrightarrow & \bar{x} \text{ is a global minimal point} \\
& \Leftrightarrow & \bar{x} \in M \text{ and } f(\bar{x}) = v \\
& \stackrel{\{x \in M \mid f(x) < v\} = \emptyset}{\Leftrightarrow} & \bar{x} \in M \text{ and } f(\bar{x}) \leq v \\
& \Leftrightarrow & \bar{x} \in \mathrm{lev}_\leq^v(f, M).
\end{array}$$

□

Exercise 1.2.16 The proof of Lemma 1.2.15 implicitly also covers the case of unsolvable problems P, i.e., $S = \emptyset$. How can one see for the various cases of unsolvability, independently of the proof of Lemma 1.2.15, that the set $\mathrm{lev}_\leq^v(f, M)$ is empty?

Lemma 1.2.17 *For some $\alpha \in \mathbb{R}$ let $\mathrm{lev}_\leq^\alpha(f, M) \neq \emptyset$. Then $S \subseteq \mathrm{lev}_\leq^\alpha(f, M)$ holds.*

Proof Because of $\mathrm{lev}_\leq^\alpha(f, M) \neq \emptyset$ there is a point $\tilde{x}$ in M with $f(\tilde{x}) \leq \alpha$. Let $\bar{x}$ be any global minimal point of P. Then $\bar{x} \in M$ and $f(\bar{x}) \leq f(\tilde{x}) \leq \alpha$ hold, so $\bar{x} \in \mathrm{lev}_\leq^\alpha(f, M)$. □

The concept of lower level sets allows us to take into account the interplay between the properties of the objective function f and of the feasible set M in a sufficient condition for solvability of P.

Theorem 1.2.18 (Strengthened Weierstrass Theorem) *For a (not necessarily bounded or closed) set $M \subseteq \mathbb{R}^n$, let $f : M \to \mathbb{R}$ be continuous, and with some $\alpha \in \mathbb{R}$ let $\text{lev}^\alpha_\leq (f, M)$ be nonempty and compact. Then S is also nonempty and compact.*

Proof Due to Lemma 1.2.17, P and the auxiliary problem

$$\widetilde{P}: \quad \min f(x) \quad \text{s.t.} \quad x \in \text{lev}^\alpha_\leq(f, M)$$

have the same optimal points and the same optimal value. $\widetilde{P}$ meets the requirements of Theorem 1.2.11, from which the assertion $S \neq \emptyset$ follows. In addition, $S = \text{lev}^v_\leq(f, M) \subseteq \text{lev}^\alpha_\leq(f, M)$ implies the boundedness of S. To prove the closedness of S, consider a convergent sequence $(x^k) \subseteq S$ with limit $x^\star$. Since S is contained in the closed set $\text{lev}^\alpha_\leq(f, M)$, it follows $x^\star \in \text{lev}^\alpha_\leq(f, M) \subseteq M$, and the continuity of f on M guarantees that from $f(x^k) \leq v, k \in \mathbb{N}$, also $f(x^\star) \leq v$ follows, so altogether $x^\star \in \text{lev}^v_\leq(f, M) = S$. Thus, S is compact. □

Exercise 1.2.19 Show that the assumptions of Theorem 1.2.18 are weaker than those of Theorem 1.2.11, that is, under the assumptions of Theorem 1.2.11 they can be fulfilled.

The *strengthening* of Theorem 1.2.18 compared to Theorem 1.2.11 refers to the statement of interest to us from the Weierstrass theorem, namely the existence of a global *minimal* point, which also follows under the weaker assumptions of Theorem 1.2.18. However, since no statement can now be made about the existence of a global *maximal* point of P, the two theorems are in fact independent of each other.

Example 1.2.20 Consider the problem

$$P: \quad \min e^x \quad \text{s.t.} \quad x \geq 0$$

(Fig. 1.17).

Here $M = \{x \in \mathbb{R} |\, x \geq 0\}$ is unbounded, so Theorem 1.2.11 is not applicable. But for example with $\alpha = e$ the set

$$\text{lev}^e_\leq(f, M) = \{x \in M |\, e^x \leq e\} = \{x \geq 0 |\, x \leq 1\} = [0, 1]$$

is nonempty and compact. Consequently, Theorem 1.2.18 is applicable and P therefore solvable.

Fig. 1.17 e^x with $x \geq 0$

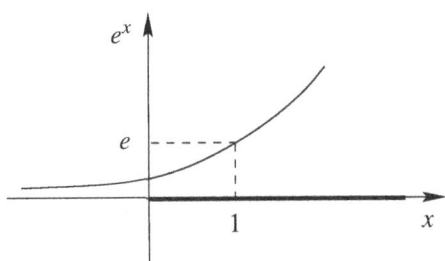

The strengthened Weierstrass theorem also shows that for the solvability of the projection problem from Example 1.1.1 the boundedness assumption on M made in Example 1.2.12 is not necessary.

Example 1.2.21 (Projection Onto a Set—Sequel 2) The projection problem from Example 1.1.1 is even solvable for every nonempty and closed set $M \subseteq \mathbb{R}^n$. To see this, we finally suppress the z-dependence of the objective function by setting $f(x) := f(x, z)$. For each $\alpha \geq 0$ the set

$$f_{\leq}^{\alpha} = \{x \in \mathbb{R}^n \mid \|x - z\|_2 \leq \alpha\}$$

then forms a ball with center z and radius α. For an arbitrary point $\tilde{x} \in M$ we choose α so large that $\tilde{x} \in f_{\leq}^{\alpha}$ holds, for example by the choice $\alpha = \|\tilde{x} - z\|_2$. Thus $\tilde{x} \in f_{\leq}^{\alpha} \cap M = \text{lev}_{\leq}^{\alpha}(f, M)$, so that $\text{lev}_{\leq}^{\alpha}(f, M)$ is nonempty, and moreover, $\text{lev}_{\leq}^{\alpha}(f, M)$ is compact as the intersection of the compact set $f_{\leq}^{\alpha}$ with the closed set M. Theorem 1.2.18 hence yields the assertion.

Corollary 1.2.22 (Strengthened Weierstrass Theorem for Unconstrained Problems) *Let the function $f : \mathbb{R}^n \to \mathbb{R}$ be continuous and, with some $\alpha \in \mathbb{R}$, let $f_{\leq}^{\alpha}$ be nonempty and compact. Then S is also nonempty and compact.*

Proof Theorem 1.2.18 with $M = \mathbb{R}^n$. □

Example 1.2.23 For $f(x) = (x-5)^2$ the set $f_{\leq}^1 = [4, 6]$ is nonempty and compact, so that f has a global minimal point on $\mathbb{R}$ according to Corollary 1.2.22.

Example 1.2.24 With $f(x) = e^x$ one obtains $f_{\leq}^{\alpha} = \emptyset$ for all $\alpha \leq 0$ and $f_{\leq}^{\alpha} = (-\infty, \log(\alpha)]$ for all $\alpha > 0$. Therefore, for no α the set $f_{\leq}^{\alpha}$ is nonempty and compact, and Corollary 1.2.22 is not applicable. Indeed, f does not possess a global minimal point on $\mathbb{R}$.

Example 1.2.25 Also for $f(x) = \sin(x)$ Corollary 1.2.22 is not applicable, as all lower level sets $f_\leq^\alpha$ are unbounded or empty. However, f does possess global minimal points on $\mathbb{R}$ (albeit with noncompact set S).

In the following we will derive a simple criterion from which the boundedness of $\text{lev}_\leq^\alpha(f, X)$ follows for *every* $\alpha \in \mathbb{R}$. This allows us to guarantee the assumptions of Theorem 1.2.18 and Corollary 1.2.22 without having to specify an explicit level α.

> **Definition 1.2.26 (Coercivity at ∞)** Consider a function $f : X \to \mathbb{R}$ with $X \subseteq \mathbb{R}^n$. If for all sequences $(x^k) \subseteq X$ with $\lim_k \|x^k\| = +\infty$ also
> $$\lim_k f(x^k) = +\infty$$
> holds, then f is called *coercive at ∞* on X. If X is a closed set, f is simply called *coercive* on X.

Example 1.2.27 The function $f(x) = (x-5)^2$ is coercive on $\mathbb{R}$.

Example 1.2.28 The function $f(x) = e^x$ is not coercive on $X = \mathbb{R}$, but it is on the set $X = \{x \in \mathbb{R} \mid x \geq 0\}$.

Example 1.2.29 (Center of a Point Cloud—Sequel 2) The objective function

$$f(z) = \left\| \begin{pmatrix} \|z - x^1\|_2 \\ \vdots \\ \|z - x^m\|_2 \end{pmatrix} \right\|_2$$

from the problem of finding the center of a point cloud (Example 1.1.5), is coercive on $\mathbb{R}^n$, because it holds

$$f(z) = \sqrt{\sum_{i=1}^m \|z - x^i\|_2^2} \geq \sqrt{\|z - x^1\|_2^2} = \|z - x^1\|_2$$

$$\geq \left| \|z\|_2 - \|x^1\|_2 \right| \xrightarrow{\|z\|_2 \to \infty} +\infty.$$

Example 1.2.30 (Cluster Analysis—Sequel 1) The objective function

$$f(z^1, \ldots, z^q) = \sum_{i=1}^m \min_{\ell=1,\ldots,q} \|z^\ell - x^i\|$$

1.2 Solvability

from problem P_1 of cluster analysis (Example 1.1.8) with $q \geq 2$ is not coercive on $\mathbb{R}^{nq}$. To see this, choose $z^{1,k} = \ldots = z^{q-1,k} = 0$, $z^{q,k} = ke_n$ (with the n-th unit vector e_n) for all $k \in \mathbb{N}$. Then we have

$$\left\| \begin{pmatrix} z^{1,k} \\ \vdots \\ z^{q,k} \end{pmatrix} \right\| \xrightarrow{k \to \infty} \infty,$$

but

$$f(z^{1,k}, \ldots, z^{q,k}) = \sum_{i=1}^{m} \min_{\ell=1,\ldots,q} \left\| z^{\ell,k} - x^i \right\|$$

$$= \sum_{i=1}^{m} \min\{ \underbrace{\min_{\ell=1,\ldots,q-1} \left\| x^i \right\|}_{\|x^i\|}, \underbrace{\left\| z^{q,k} - x^i \right\|}_{\geq \| \| z^{q,k} \| - \| x^i \| \|} \} = \sum_{i=1}^{m} \left\| x^i \right\|$$

for all sufficiently large $k \in \mathbb{N}$. Since the last expression is independent of $(z^{1,k}, \ldots, z^{q,k})$ and therefore does not tend to infinity for $k \to \infty$, f is not coercive on any set that contains the sequence of points $(z^{1,k}, \ldots, z^{q,k})$, and thus in particular not on $\mathbb{R}^{nq}$.

Example 1.2.31 On compact sets X, every function f is trivially coercive.

Note that the term 'trivial' is used carefully in this textbook. It does not refer to statements that are 'easy' to prove from the author's point of view, but to those that are valid due to a logical triviality. For example, the statement 'All unicorns are pink' is trivially true, because to refute it, one would have to find a unicorn that is not pink. But (spoiler alert!) since one cannot find a unicorn in the first place, there is no need to look for a unicorn that is not pink. Therefore, the statement cannot be refuted for a trivial reason and is consequently true. In Example 1.2.31, this refers to the fact that in a compact set X, there is not a single sequence (x^k) with $\lim_k \|x^k\| \to +\infty$. To show that f is *not* coercive, however, such a sequence would have to *exist* and also fail to satisfy $\lim_k f(x^k) = +\infty$. The latter is irrelevant, however, because the existence of the sequence is not given. Consequently, f is coercive on X for a trivial reason. For the following result, note that the empty set is *trivially* bounded in the same sense.

Lemma 1.2.32 *Let the function $f : X \to \mathbb{R}$ be coercive at ∞ on the (not necessarily bounded or closed) set $X \subseteq \mathbb{R}^n$. Then the set $\mathrm{lev}_{\leq}^{\alpha}(f, X)$ is bounded for every level $\alpha \in \mathbb{R}$.*

Proof We choose an arbitrary $\alpha \in \mathbb{R}$ and assume that $\mathrm{lev}_{\leq}^{\alpha}(f, X)$ is unbounded. Then there exists a sequence $(x^k) \subseteq \mathrm{lev}_{\leq}^{\alpha}(f, X)$ with $\lim_k \|x^k\| = +\infty$. Due to the coercivity at ∞ of f on X, this implies $\lim_k f(x^k) = +\infty$. On the other hand, by the choice of (x^k) we have $f(x^k) \leq \alpha$ for all $k \in \mathbb{N}$, a contradiction. Therefore, $\mathrm{lev}_{\leq}^{\alpha}(f, X)$ is bounded. □

In the Weierstrass theorem, we can replace the boundedness of M by the coercivity of f on M in the sense of the following corollary.

> **Corollary 1.2.33** Let M be nonempty and closed, but not necessarily bounded. Furthermore, let the function $f : M \to \mathbb{R}$ be continuous and coercive on M. Then S is nonempty and compact.

Proof In view of $M \neq \emptyset$ we can choose some $\bar{x} \in M$ and define $\alpha = f(\bar{x})$. Then the set $\text{lev}_{\leq}^{\alpha}(f, M)$ contains the point $\bar{x}$ and is therefore nonempty.

Due to the continuity of f on M and the closedness of M, $\text{lev}_{\leq}^{\alpha}(f, M)$ is closed and, according to Lemma 1.2.32, also bounded. Overall, we have found an $\alpha \in \mathbb{R}$ such that $\text{lev}_{\leq}^{\alpha}(f, M)$ is nonempty and compact. Thus, Theorem 1.2.18 provides the assertion. □

Example 1.2.34 (Center of a Point Cloud—Sequel 3) The problem of finding the center of a point cloud (Example 1.1.5), is solvable according to Example 1.2.29, the continuity of its objective function, and Corollary 1.2.33.

Because of Example 1.2.30, Corollary 1.2.33 cannot be used to demonstrate the solvability of the problem P_1 of cluster analysis.

In the following we will show that P_1 is nevertheless solvable, using an alternative way to apply the Weierstrass theorem (Theorem 1.2.11) to unconstrained problems.

Example 1.2.35 (Cluster Analysis—Sequel 2) To show that the problem P_1 of cluster analysis (Example 1.1.8) with the Euclidean norm as inner norm, i.e., the unconstrained minimization of the objective function

$$f(z^1, \ldots, z^q) = \left\| \begin{pmatrix} \min_{\ell=1,\ldots,q} \| z^\ell - x^1 \|_2 \\ \vdots \\ \min_{\ell=1,\ldots,q} \| z^\ell - x^m \|_2 \end{pmatrix} \right\|_1$$

for $q \geq 2$, is solvable, we will construct a nonempty and compact set $M \subseteq \mathbb{R}^{nq}$, outside of which one does not need to search for global minimal points. The unconstrained minimization of f is therefore equivalent to the minimization of f over M, and Theorem 1.2.11 provides the assertion.

For this, we consider the box (for this term see also Sect. 3.3) $X = [\underline{x}_1, \bar{x}_1] \times \ldots \times [\underline{x}_n, \bar{x}_n]$ with

$$\underline{x}_j := \min_{i=1,\ldots,m} x_j^i, \quad \bar{x}_j := \max_{i=1,\ldots,m} x_j^i, \quad j = 1, \ldots, n,$$

i.e., the smallest box in $\mathbb{R}^n$ that contains all data points, and set $M := \prod_{\ell=1}^q X$.

In the following, we will show that for every point $(z^1, \ldots, z^q)$ in $\mathbb{R}^{nq}$ there exists a point $(\tilde{z}^1, \ldots, \tilde{z}^q)$ in M with at least as good objective function value. Even if f should have global minimal points outside of M, there would be other global minimal points in M, so that the constrained minimization of f over M instead of the unconstrained minimization provides a correct result.

1.2 Solvability

So let $(z^1, \ldots, z^q) \in \mathbb{R}^{nq}$. We set for each $\ell \in \{1, \ldots, q\}$ and each $j \in \{1, \ldots, n\}$

$$\tilde{z}_j^\ell = \begin{cases} z_j^\ell, & \text{if } z_j^\ell \in [\underline{x}_j, \overline{x}_j] \\ \underline{x}_j, & \text{if } z_j^\ell < \underline{x}_j \\ \overline{x}_j, & \text{if } z_j^\ell > \overline{x}_j. \end{cases}$$

Then the point $(\tilde{z}^1, \ldots, \tilde{z}^q)$ lies in M. Furthermore, for each $\ell \in \{1, \ldots, q\}, j \in \{1, \ldots, n\}$ and $i \in \{1, \ldots, m\}$ in the first case above

$$|\tilde{z}_j^\ell - x_j^i| = |z_j^\ell - x_j^i|,$$

in the second case

$$|\tilde{z}_j^\ell - x_j^i| = |\underline{x}_j - x_j^i| = x_j^i - \underline{x}_j < x_j^i - z_j^\ell = |z_j^\ell - x_j^i|$$

and in the third case

$$|\tilde{z}_j^\ell - x_j^i| = |\overline{x}_j - x_j^i| = \overline{x}_j - x_j^i < z_j^\ell - x_j^i = |z_j^\ell - x_j^i|$$

hold, so in each of the three cases $|\tilde{z}_j^\ell - x_j^i| \leq |z_j^\ell - x_j^i|$. With the definition of the Euclidean norm, it is easy to see that then also

$$\|\tilde{z}^\ell - x^i\|_2 \leq \|z^\ell - x^i\|_2 \tag{1.1}$$

is true. From this follows

$$\min_{\ell=1,\ldots,q} \|\tilde{z}^\ell - x^i\|_2 \leq \min_{\ell=1,\ldots,q} \|z^\ell - x^i\|_2$$

and, by definition of the ℓ_1-norm,

$$f(\tilde{z}^1, \ldots, \tilde{z}^q) \leq f(z^1, \ldots, z^q). \tag{1.2}$$

This is the assertion.

We note that in the above example, to prove both (1.1) and (1.2), we used the property of the involved norms that $|a_i| \leq |b_i|, i = 1, \ldots, n$, implies $\|a\| \leq \|b\|$. Every norm with this property is called *abs-monotone* on $\mathbb{R}^n$ (this property is often referred to as monotonicity in the literature, but this is not consistent with the definition of monotonicity for functionals; Definition 1.3.8).

Exercise 1.2.36 Show that all ℓ_p-norms with $p \in [1, \infty]$ are abs-monotone on $\mathbb{R}^n$.

Exercise 1.2.37 Provide an example of a norm on $\mathbb{R}^2$ that is not abs-monotone.

Since apart from the abs-monotonicity of the involved norms no further special properties were used, the arguments from Example 1.2.35 transfer directly to the solvability of problems of cluster analysis with any inner and outer norms, as long as both are abs-monotone (e.g., as ℓ_p-norms; Exercise 1.2.36).

Furthermore, with the same idea an alternative proof for the solvability of the problem to determine the center of a point cloud can be given, which then is also transferable to all abs-monotone inner and outer norms.

Exercise 1.2.38 On the closed set $X \subseteq \mathbb{R}^n$ let the functions f, g_i, $i \in I$, be continuous, let the set $M := \{x \in X | g_i(x) \leq 0, i \in I\}$ be nonempty, and let at least one of the functions f, g_i, $i \in I$, be coercive on X. Show that the set S of optimal points of f on M is then nonempty and compact.

1.2.5 Nonclosed Feasible Sets

The following example shows that also optimization problems with nonclosed feasible sets can occur in applications. Their solvability cannot be guaranteed by the results derived so far.

Example 1.2.39 (Maximum Likelihood Estimator) Let N observations $\widehat{x}_1, \ldots, \widehat{x}_N \geq 0$ with $\bar{x} = \frac{1}{N}\sum_{i=1}^{N} \widehat{x}_i > 0$ be given, which are considered as realizations of stochastically independent and with parameter $\lambda > 0$ exponentially distributed random variables $X_1, \ldots, X_N$. The parameter λ that 'fits best' to the observations is sought. For this purpose, using the density functions of the individual X_i,

$$f(\lambda, x_i) = \begin{cases} \lambda e^{-\lambda x_i}, & x_i \geq 0 \\ 0, & x_i < 0, \end{cases}$$

first consider the joint density of all random variables

$$L(\lambda, x) = \prod_{i=1}^{N} f(\lambda, x_i).$$

The maximum likelihood estimator then determines λ as the (unique) optimal point of the problem

$$ML: \quad \max_{\lambda} L(\lambda, \widehat{x}) \quad \text{s.t.} \quad \lambda > 0.$$

The feasible set $M = (0, +\infty)$ of this problem is not closed. It is also pointless to artificially add the parameter value $\lambda = 0$, as $f(0, x)$ is not a probability density.

We will see how the solvability of this problem can still be guaranteed, and later also determine the unique global maximal point. For this, we first calculate

$$L(\lambda, \widehat{x}) = \prod_{i=1}^{N} \lambda e^{-\lambda \widehat{x}_i} = \lambda^N e^{-\lambda N \bar{x}}.$$

Figure 1.18 shows the graph of this function in the case $N = 2$ and $\bar{x} = 1$.

Since many optimization algorithms require the computation of at least first derivatives, and since the likelihood function L has a pronounced product structure

1.2 Solvability

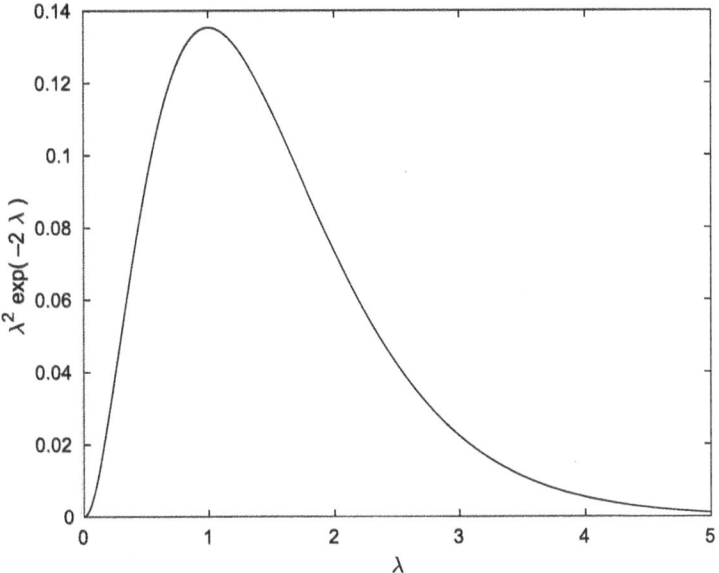

Fig. 1.18 Graph of a likelihood function L

and positive values everywhere, it seems advisable to replace it by the *log-likelihood function*

$$\ell(\lambda, \widehat{x}) := \log(L(\lambda, \widehat{x})) = N \log(\lambda) - \lambda N \bar{x}.$$

Indeed, since the log function is strictly monotonically increasing on $(0, +\infty)$ and therefore on the image set of L, one can show by means of Exercise 1.3.5 that the problem

$$ML_{\log}: \quad \max_{\lambda} \ell(\lambda, \widehat{x}) \quad \text{s.t.} \quad \lambda > 0$$

has the same optimal points as *ML*. Finally, we eliminate the constant $N > 0$ from the objective function using Exercise 1.3.1a and move to the equivalent minimization problem

$$P_{ML}: \quad \min_{\lambda} \lambda \bar{x} - \log(\lambda) \quad \text{s.t.} \quad \lambda > 0,$$

whose objective function for $\bar{x} = 1$ is plotted in Fig. 1.19.

Example 1.2.39 motivates that one should require the coercivity of a function f on a *not necessarily closed* set X not only 'at ∞' as in Definition 1.2.26, but also at certain 'finite points'.

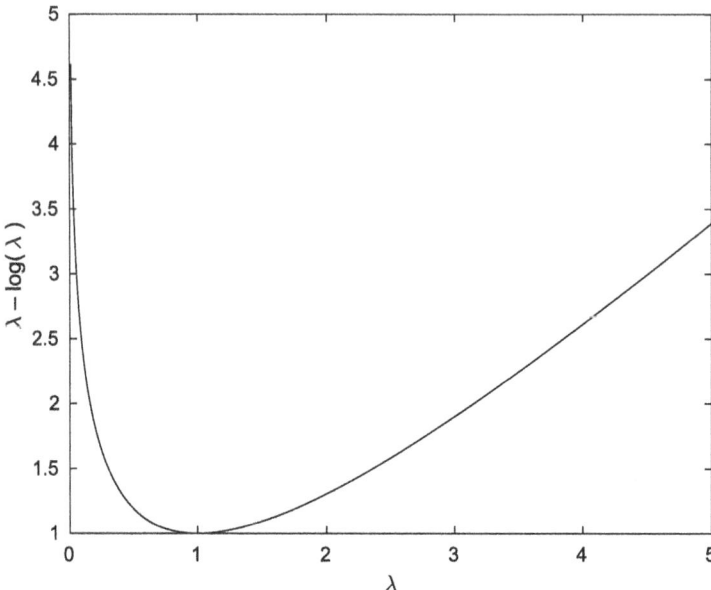

Fig. 1.19 Objective function of the problem P_{ML}

Definition 1.2.40 (Coercivity) Consider a (not necessarily closed) set $X \subseteq \mathbb{R}^n$ and a function $f : X \to \mathbb{R}$. If for all sequences $(x^k) \subseteq X$ with $\lim_k \|x^k\| = +\infty$ and all convergent sequences $(x^k) \subseteq X$ with $\lim_k x^k \notin X$ the condition

$$\lim_k f(x^k) = +\infty$$

holds, then f is called *coercive* on X.

For closed sets X, the concept of coercivity from Definition 1.2.40 agrees with the coercivity at ∞ from Definition 1.2.26, which justifies the shorthand introduced there for coercivity at ∞ on closed sets.

Example 1.2.41 (Maximum Likelihood Estimator—Sequel 1) For $\bar{x} > 0$, the objective function $f(\lambda) = \lambda \bar{x} - \log(\lambda)$ of the problem P_{ML} from Example 1.2.39 is coercive on $M = (0, +\infty)$. We note that only the log-likelihood function ℓ leads to coercivity, while the likelihood function L itself does *not*.

1.3 Rules of Calculus and Transformations

Lemma 1.2.42 *Let the function $f : X \to \mathbb{R}$ be continuous and coercive on the (not necessarily bounded or closed) set $X \subseteq \mathbb{R}^n$. Then the set $\mathrm{lev}_{\leq}^{\alpha}(f, X)$ is compact for every level $\alpha \in \mathbb{R}$.*

Proof We choose some arbitrary $\alpha \in \mathbb{R}$. According to Lemma 1.2.32, the set $\mathrm{lev}_{\leq}^{\alpha}(f, X)$ is bounded. Its closedness is trivial in the case $\mathrm{lev}_{\leq}^{\alpha}(f, X) = \emptyset$. Otherwise we choose a convergent sequence $(x^k) \subseteq \mathrm{lev}_{\leq}^{\alpha}(f, X)$, whose limit point we denote by $x^{\star}$. Suppose that $x^{\star}$ does not lie in X. Due to the coercivity of f on X it would then follow $\lim_k f(x^k) = +\infty$, in contradiction to $f(x^k) \leq \alpha$ for all $k \in \mathbb{N}$. Therefore, $x^{\star} \in X$ holds, and the continuity of f at $x^{\star}$ finally yields $f(x^{\star}) \leq \alpha$, so overall $x^{\star} \in \mathrm{lev}_{\leq}^{\alpha}(f, X)$. Thus, $\mathrm{lev}_{\leq}^{\alpha}(f, X)$ is also closed and therefore compact. □

Through the above extension of the concept of coercivity, we can replace not only the boundedness (as in Corollary 1.2.33), but also the closedness of M by the coercivity of f on M in the Weierstrass theorem in the sense of the following corollary.

Corollary 1.2.43 *Let M be nonempty, but not necessarily bounded or closed. Furthermore, let the function $f : M \to \mathbb{R}$ be continuous and coercive on M. Then, S is nonempty and compact.*

Proof As in the proof of Corollary 1.2.33, we set $\alpha = f(\bar{x})$ with some $\bar{x} \in M$ and obtain the consistency of the set $\mathrm{lev}_{\leq}^{\alpha}(f, M)$. According to Lemma 1.2.42, this set is also compact, so we have again found some $\alpha \in \mathbb{R}$ such that $\mathrm{lev}_{\leq}^{\alpha}(f, M)$ is nonempty and compact. Thus, Theorem 1.2.18 provides the assertion. □

Example 1.2.44 (Maximum Likelihood Estimator—Sequel 2) The problem P_{ML} from Example 1.2.39 is solvable according to Example 1.2.41, the continuity of its objective function, and Corollary 1.2.43. Since both problems possess the same set of optimal points, also the original maximum likelihood problem ML is solvable.

1.3 Rules of Calculus and Transformations

This section introduces a series of rules of calculus and transformations of optimization problems that are of interest in this textbook. The existence of all occurring optimal points and values is assumed in this section without further mention and must be guaranteed when applying the results, for example, using the techniques from Sect. 1.2. The transfer of the results on optimal values to cases of nonattained infima and suprema is left to the reader as further exercise.

Exercise 1.3.1 (Scalar Multiples and Sums) Let $M \subseteq \mathbb{R}^n$ and $f, g : M \to \mathbb{R}$. Then the following holds:

(a) $\forall \alpha \geq 0, \beta \in \mathbb{R}: \min_{x \in M} (\alpha f(x) + \beta) = \alpha (\min_{x \in M} f(x)) + \beta.$
(b) $\forall \alpha < 0, \beta \in \mathbb{R}: \min_{x \in M} (\alpha f(x) + \beta) = \alpha (\max_{x \in M} f(x)) + \beta.$
(c) $\min_{x \in M} (f(x) + g(x)) \geq \min_{x \in M} f(x) + \min_{x \in M} g(x).$
(d) In statement c, the strict inequality $>$ can occur.

In statement a and statement b, the local and global optimal points, respectively, of the optimization problems also coincide.

Exercise 1.3.2 (Separable Objective Function on Cartesian Product) Let $X \subseteq \mathbb{R}^n$, $Y \subseteq \mathbb{R}^m$, $f : X \to \mathbb{R}$ and $g : Y \to \mathbb{R}$. Then it holds

$$\min_{(x,y) \in X \times Y} (f(x) + g(y)) = \min_{x \in X} f(x) + \min_{y \in Y} g(y).$$

Exercise 1.3.3 (Swapping Minimizations and Maximizations) Let $X \subseteq \mathbb{R}^n$, $Y \subseteq \mathbb{R}^m$, $M = X \times Y$ and $f : M \to \mathbb{R}$. Then it holds:

(a) $\min_{(x,y) \in M} f(x, y) = \min_{x \in X} \min_{y \in Y} f(x, y) = \min_{y \in Y} \min_{x \in X} f(x, y).$
(b) $\max_{(x,y) \in M} f(x, y) = \max_{x \in X} \max_{y \in Y} f(x, y) = \max_{y \in Y} \max_{x \in X} f(x, y).$
(c) $\min_{x \in X} \max_{y \in Y} f(x, y) \geq \max_{y \in Y} \min_{x \in X} f(x, y).$
(d) In statement c, the strict inequality $>$ can occur.

Exercise 1.3.4 (Union) Let I be an arbitrary index set, $M_i \subseteq \mathbb{R}^n$, $i \in I$, and $f : \bigcup_{i \in I} M_i \to \mathbb{R}$. Then it holds

$$\min_{x \in \bigcup_{i \in I} M_i} f(x) = \min_{i \in I} \min_{x \in M_i} f(x).$$

Exercise 1.3.5 (Monotone Transformation) For $M \subseteq \mathbb{R}^n$ and a function $f : M \to Y$ with $Y \subseteq \mathbb{R}$, let $\psi : Y \to \mathbb{R}$ be a strictly monotonically increasing function. Then it holds

$$\min_{x \in M} \psi(f(x)) = \psi (\min_{x \in M} f(x)),$$

and the local and global minimal points, respectively, coincide.

1.3 Rules of Calculus and Transformations

For the following exercise recall the definition of a parallel projection from Definition 1.2.8.

Exercise 1.3.6 (Projection Reformulation) Consider $M \subseteq \mathbb{R}^n \times \mathbb{R}^m$ and a function $f : \mathbb{R}^n \to \mathbb{R}$, which does not depend on the variables from $\mathbb{R}^m$. Then the problems

$$P: \min_{(x,y) \in \mathbb{R}^n \times \mathbb{R}^m} f(x) \quad \text{s.t.} \quad (x, y) \in M$$

and

$$P_{\text{proj}}: \min_{x \in \mathbb{R}^n} f(x) \quad \text{s.t.} \quad x \in \text{pr}_x M$$

are equivalent in the following sense:

(a) For every local or global minimal point $(x^\star, y^\star)$ of P, $x^\star$ is a local or global minimal point of P_{proj}, respectively.
(b) For every local or global minimal point $x^\star$ of P_{proj}, there exists some $y^\star \in \mathbb{R}^m$ such that $(x^\star, y^\star)$ is a local or global minimal point of P, respectively.
(c) The minimal values of P and P_{proj} coincide.

Exercise 1.3.7 (Epigraph Reformulation) Consider $M \subseteq \mathbb{R}^n$ and a function $f : M \to \mathbb{R}$. Then the problems

$$P: \min_{x \in \mathbb{R}^n} f(x) \quad \text{s.t.} \quad x \in M$$

and

$$P_{\text{epi}}: \min_{(x,\alpha) \in \mathbb{R}^n \times \mathbb{R}} \alpha \quad \text{s.t.} \quad f(x) \leq \alpha, \; x \in M$$

are equivalent in the following sense:

(a) For every local or global minimal point $x^\star$ of P, $(x^\star, f(x^\star))$ is a local or global minimal point of P_{epi}, respectively.
(b) For every local or global minimal point $(x^\star, \alpha^\star)$ of P_{epi}, $x^\star$ is a local or global minimal point of P, respectively.
(c) The minimal values of P and P_{epi} coincide.

Definition 1.3.8 (Monotone Functional)
We call $F : \mathbb{R}^k \to \mathbb{R}$ *monotone* (on $\mathbb{R}^k$), if

$$\forall x, y \in \mathbb{R}^k \text{ with } x \leq y : \quad F(x) \leq F(y)$$

applies. The inequalities between vectors are to be understood componentwise.

Exercise 1.3.9 (Generalized Epigraph Reformulation) Consider $X \subseteq \mathbb{R}^n$, functions $f : X \to \mathbb{R}^k$ and $g : X \to \mathbb{R}^\ell$ as well as monotone functionals $F : \mathbb{R}^k \to \mathbb{R}$ and $G : \mathbb{R}^\ell \to \mathbb{R}$. Then the problems

$$P : \min_{x \in \mathbb{R}^n} F(f(x)) \quad \text{s.t.} \quad G(g(x)) \leq 0, \quad x \in X$$

and

$$P_{\text{epi}} : \min_{(x,\alpha,\beta) \in \mathbb{R}^n \times \mathbb{R}^k \times \mathbb{R}^\ell} F(\alpha) \quad \text{s.t.} \quad G(\beta) \leq 0,$$
$$f(x) \leq \alpha,$$
$$g(x) \leq \beta,$$
$$x \in X$$

are equivalent in the following sense:

(a) For every local or global minimal point $x^\star$ of P, $(x^\star, f(x^\star), g(x^\star))$ is a local or global minimal point of P_{epi}, respectively.
(b) For every local or global minimal point $(x^\star, \alpha^\star, \beta^\star)$ of P_{epi}, $x^\star$ is a local or global minimal point of P, respectively.
(c) The minimal values of P and P_{epi} coincide.

Exercise 1.3.10 Formulate an equivalent linear optimization problem to the nonsmooth optimization problem

$$P : \min_{x \in \mathbb{R}^2} \ (\max\{x_1 + 3x_2, -x_1 + x_2\} + 2\max\{5x_1 - x_2, -3x_1 + x_2, x_1\})$$
$$\text{s.t.} \quad x_1 - x_2 + \max\{x_1 + 7x_2, 2x_1 - x_2\} + \max\{-x_1 - x_2, x_1 + 4x_2\} \leq 0.$$

Convex Optimization Problems

Contents

2.1 Convexity 36
2.2 The C^1-Characterization of Convexity 41
    2.2.1 Multidimensional First Derivatives 41
    2.2.2 C^1-Characterization 44
2.3 Solvability of Convex Optimization Problems 46
2.4 Optimality Conditions for Unconstrained Convex Problems 47
2.5 The C^2-Characterization of Convexity 51
    2.5.1 The Multidimensional Second Derivative 51
    2.5.2 C^2-Characterizations 53
2.6 The Monotonicity Characterization of Convexity 57
2.7 Optimality Conditions for Constrained Convex Problems 58
    2.7.1 Lagrange and Wolfe Duality 59
    2.7.2 The Karush-Kuhn-Tucker Conditions 72
    2.7.3 Complementarity 75
    2.7.4 Geometric Interpretation of the KKT Conditions 76
    2.7.5 Constraint Qualifications 78
2.8 Algorithms 87
    2.8.1 Basic Idea of the Gradient Method 88
    2.8.2 Basic Idea of the Newton Method 89
    2.8.3 Basic Idea of Cutting Plane Methods 90
    2.8.4 Kelley's Cutting Plane Method... 92
    2.8.5 The Frank-Wolfe Method 98
    2.8.6 Basic Idea of Primal-Dual Interior Point Methods 102

For minimization problems with a convex feasible set and a convex objective function, in addition to strong theoretical results, effective algorithms can often be formulated. In the context of this textbook, we consider convexity only for sufficiently often continuously differentiable functions, which leads to particularly transparent results without overly restricting the relevance to applications. To transfer these results to nonsmooth functions, see, for example, [34].

After an introduction to the basics of convex sets and functions in Sect. 2.1, we use the additional smoothness requirement for convex functions in Sect. 2.2 and prove the C^1-characterization of convexity. Although convexity does not provide decisive advantages for the solvability of optimization problems, we provide a partial result in Sect. 2.3. Also based on the C^1-characterization of convexity, we characterize the set of global minimal points of *unconstrained* optimization problems by the solution of an equation in Sect. 2.4.

In the subsequent Sect. 2.5, we derive a handy way to check the convexity of functions by the C^2-characterization of convexity. For completeness, we also provide a characterization of convexity by monotonicity of the first derivative in Sect. 2.6. Optimality conditions and estimates of the optimal value for *constrained* convex optimization problems are based on duality statements, which we prove in Sect. 2.7. Section 2.8 then discusses a series of algorithmic approaches for convex optimization problems, the most effective of which also rely on the explicit exploitation of duality.

2.1 Convexity

Definition 2.1.1 (Convex Sets and Functions)

(a) A set $X \subseteq \mathbb{R}^n$ is called *convex*, if

$$\forall x, y \in X, \lambda \in (0,1): \quad (1-\lambda)x + \lambda y \in X$$

holds (i.e., the connecting line between every pair of two points in X belongs entirely to X; Fig. 2.1).

(b) For a convex set $X \subseteq \mathbb{R}^n$ a function $f : X \to \mathbb{R}$ is called *convex (on X)*, if

$$\forall x, y \in X, \lambda \in (0,1): \quad f((1-\lambda)x + \lambda y) \leq (1-\lambda)f(x) + \lambda f(y)$$

holds (i.e., the graph of the function f lies below each of its secants; Fig. 2.2).

(c) For a convex set $X \subseteq \mathbb{R}^n$ a function $f : X \to \mathbb{R}$ is called *strictly convex (on X)*, if in part b for $x \neq y$ even the strict inequality < holds (i.e., the graph of the function f lies properly below each of its secants).

(d) For a convex set $X \subseteq \mathbb{R}^n$ a function $f : X \to \mathbb{R}$ is called *strongly convex (on X)*, if with some constant $c > 0$ the function $f(x) - \frac{c}{2}\|x\|_2^2$ is convex on X.

(e) For a convex set $X \subseteq \mathbb{R}^n$ a function $f : X \to \mathbb{R}$ is called *concave, strictly concave* or *strongly concave (on X)*, if $-f$ is convex, strictly convex or strongly convex on X, respectively.

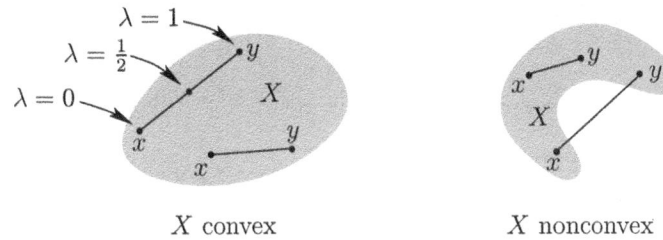

Fig. 2.1 Convexity of sets in $\mathbb{R}^2$

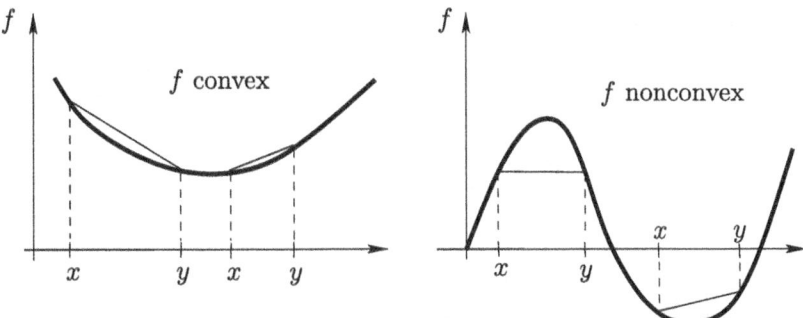

Fig. 2.2 Convexity of functions on $\mathbb{R}$

The following examples illustrate the concept of convexity for sets and functions.

- The sets ∅ and $\mathbb{R}^n$ are convex.
- The set $\{x \in \mathbb{R}^2 \mid x_1 \geq 0\}$ is convex (convex sets thus do not need to be bounded).
- The set $\{x \in \mathbb{R}^2 \mid x_1^2 + x_2^2 < 1\}$ is convex (convex sets thus do not need to be closed).
- The function $f(x) = \sin(x)$ is concave on $X_1 = [0, \pi]$, convex on $X_2 = [\pi, 2\pi]$ and neither convex nor concave on $X_3 = [0, 2\pi]$.
- The function $f(x) = |x|$ is convex on $\mathbb{R}$, and the function $f(x) = -\sqrt{1-x^2}$ is convex on $[-1, 1]$ (convex functions thus do not need to be differentiable).
- Every affine-linear function $f(x) = a^\mathsf{T} x + b$ with $a \in \mathbb{R}^n$ and $b \in \mathbb{R}$ is convex, but not strictly convex.

In the following, we will often loosely refer to affine-linear functions as linear. Strictly speaking, this means $b = 0$, but such a distinction is irrelevant within the scope of this textbook.

- The function $f(x) = (x-5)^2$ is strongly and thus also strictly convex on $\mathbb{R}$ (Example 2.5.11).
- The function $f(x) = e^x$ is strictly convex, but not strongly convex on $\mathbb{R}$ (Example 2.5.12).

- The function $f(z) = \|(\|z - x^1\|_2, \ldots, \|z - x^m\|_2)^\mathsf{T}\|_2$ from Example 1.1.5 is convex on $\mathbb{R}^n$.
- The function $f(z^1, \ldots, z^q) = \sum_{i=1}^{m} \min_{\ell=1,\ldots,q} \|z^\ell - x^i\|$ with $q \geq 2$ from Example 1.1.8 is not convex on $\mathbb{R}^{nq}$.
- The concept of a 'concave set' does *not* exist.

Exercise 2.1.2 Show the equivalence of the following statements:

(a) The set $X \subseteq \mathbb{R}^n$ and the function $f : X \to \mathbb{R}$ are convex.
(b) The set $\mathrm{epi}(f, X)$ is convex.

Exercise 2.1.3 On a convex set $X \subseteq \mathbb{R}^n$ let the functions $f, g : X \to \mathbb{R}$ be convex. Show that then for all $\sigma, \mu \geq 0$ the function $\sigma f + \mu g$ is convex on X.

Exercise 2.1.4 Show that for every norm $\|\cdot\|$ on $\mathbb{R}^n$ the function $f(x) = \|x\|$ is convex on $\mathbb{R}^n$.

> **Definition 2.1.5 (Convex Optimization Problem)** The optimization problem
> $$P: \quad \min f(x) \quad \text{s.t.} \quad x \in M$$
> is called *convex*, if the set M and the function $f : M \to \mathbb{R}$ are convex.

Since $M = \mathbb{R}^n$ is a convex set, an unconstrained problem is convex if and only if f is convex on $\mathbb{R}^n$.

The following theorem is of central importance for solving convex optimization problems.

> **Theorem 2.1.6** *Let the optimization problem P be convex. Then every local minimal point of P is also a global minimal point of P.*

Proof Let the point $\bar{x} \in M$ be a local minimal point of P. Suppose, $\bar{x}$ is not a global minimal point of P. Then there exists some $y \in M$ with $f(y) < f(\bar{x})$. The points on the connecting line between $\bar{x}$ and y, i.e., $x(\lambda) = (1 - \lambda)\bar{x} + \lambda y$ with $\lambda \in (0, 1)$, lie all in M due to the convexity of M, and because of the convexity of f on M we have for all $\lambda \in (0, 1)$

$$f(x(\lambda)) \leq (1 - \lambda)f(\bar{x}) + \lambda \underbrace{f(y)}_{< f(\bar{x})} < f(\bar{x}).$$

2.1 Convexity

Fig. 2.3 Proof idea for Theorem 2.1.6

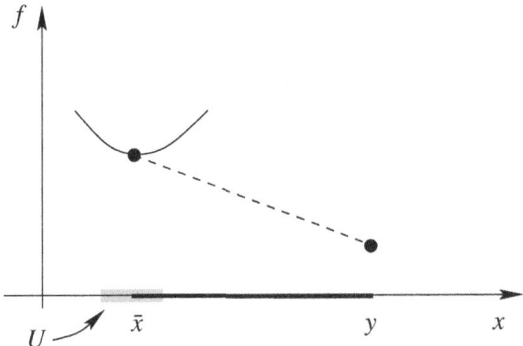

Consequently, for every neighborhood U of $\bar{x}$ there exists some $\lambda \in (0, 1)$ with $x(\lambda) \in U \cap M$ and $f(x(\lambda)) < f(\bar{x})$ (Fig. 2.3). However, this contradicts the fact that $\bar{x}$ is a local minimal point of P. Therefore, the assumption that $\bar{x}$ is not a global minimal point is wrong. $\square$

In convex optimization problems P, it is therefore sufficient to only search for local minimal points to find global minimal points. The reason for this effect is that convexity is a *global* requirement for P.

In applications, the feasible set M is often not given abstractly, but is described by equality and inequality constraints. In the following, we derive properties of the involved functions that guarantee the convexity of M.

Exercise 2.1.7 Let the set $X \subseteq \mathbb{R}^n$ and the function $f : X \to \mathbb{R}$ be convex. Show that then for each $\alpha \in \overline{\mathbb{R}}$ the lower level set $\mathrm{lev}^\alpha_\leq(f, X)$ is convex. Also show that the reverse of this statement is false.

Exercise 2.1.8 Show that the intersection of any number of convex sets is again convex.

Lemma 2.1.9 *For any index set I let the functions $g_i : \mathbb{R}^n \to \mathbb{R}$, $i \in I$, be convex. Then the set $M = \{x \in \mathbb{R}^n | g_i(x) \leq 0, i \in I\}$ is convex.*

Proof From the representation

$$M = \bigcap_{i \in I} (g_i)^0_\leq$$

the claim follows with Exercises 2.1.7 and 2.1.8. $\square$

Next we clarify when a set $H = \{x \in \mathbb{R}^n | h(x) = 0\}$ described by an *equality* constraint with $h : \mathbb{R}^n \to \mathbb{R}$ is convex. The example $h(x) = x_1^2 + x_2^2 - 1$ shows that convexity of h is *not* sufficient. Due to

$$h(x) = 0 \iff h(x) \leq 0 \text{ and } -h(x) \leq 0$$

it follows with Exercise 2.1.7 and 2.1.8 that H is convex if both h and $-h$ are convex. This is equivalent to h being both convex and concave, so

$$\forall\, x, y \in \mathbb{R}^n,\, \lambda \in (0, 1) : \quad h((1-\lambda)x + \lambda y) = (1-\lambda)h(x) + \lambda h(y)$$

holds. Consequently, for every linear function h the set H is convex.

Definition 2.1.10 (Convexly Described Set) We call a set given by inequality and equality constraints with arbitrary index sets I and J,

$$M = \{x \in \mathbb{R}^n | g_i(x) \leq 0,\, i \in I,\, h_j(x) = 0,\, j \in J\},$$

convexly described, if the functions $g_i : \mathbb{R}^n \to \mathbb{R}$, $i \in I$, are convex and the functions $h_j : \mathbb{R}^n \to \mathbb{R}$, $j \in J$, are linear.

In this terminology, our above considerations yield the following result.

Lemma 2.1.11 *Every convexly described set is convex.*

Example 2.1.12 If f, g_i, $i \in I$, are convex on $\mathbb{R}^n$ and h_j, $j \in J$, are linear functions, then

$$P : \quad \min f(x) \quad \text{s.t.} \quad g_i(x) \leq 0,\, i \in I,\, h_j(x) = 0,\, j \in J,$$

is a convex optimization problem. In this case, we call P a *convexly described* optimization problem.

Example 2.1.13 The sufficient condition for convexity of M from Lemma 2.1.11 is *not* necessary, i.e., there are convex sets that are not convexly described. This can be seen, for example, by the epigraph reformulation of the problem considered in Example 1.1.7 to determine the center of a point cloud with maximum norm as the outer norm: The constraint functions $g_i(z, \alpha) := \|z - x^i\|_2 - \alpha$ associated with the constraints $\|z - x^i\|_2 \leq \alpha$, $i = 1, \ldots, m$, are convex on $\mathbb{R}^n \times \mathbb{R}$, so that according to Lemma 2.1.11 the feasible set of the problem P_{epi} is convex. However, if one wants to avoid the nondifferentiability of the functions g_i through the equivalence transformation

$$\|z - x^i\|_2 \leq \alpha \quad \iff \quad \|z - x^i\|_2^2 \leq \alpha^2,\, \alpha \geq 0$$

then the new constraint functions $G_i(z, \alpha) := \|z - x^i\|_2^2 - \alpha^2$, $i = 1, \ldots, m$, are differentiable, but no longer convex (due to the term $-\alpha^2$). The equivalent transformation has not changed the *geometry* of the feasible set. It is therefore still convex, but is now described by nonconvex inequality constraints.

In such cases, where sets, although not convexly described, are convex (or even if only an equivalent convex description exists), one speaks of *hidden convexity*.

Example 2.1.14 (Linear Optimization) With $c \in \mathbb{R}^n$, $b \in \mathbb{R}^m$ and an (m, n)-matrix

$$A = \begin{pmatrix} a_1^T \\ \vdots \\ a_m^T \end{pmatrix}$$

with $a_i \in \mathbb{R}^n$, $i = 1, \ldots, m$,

$$P: \quad \min c^T x \quad \text{s.t.} \quad Ax \leq b$$

is a linear optimization problem. If desired, the inequality constraints can absorb some not explicitly stated nonnegativity constraints $x \geq 0$, and also model equality constraints.

P is also a *convex* (and even convexly described) optimization problem, because by setting $f(x) = c^T x$, $I = \{1, \ldots, m\}$ and $g_i(x) = a_i^T x - b_i$, $i \in I$, the functions $f, g_i : \mathbb{R}^n \to \mathbb{R}$, $i \in I$, are linear and therefore convex on $\mathbb{R}^n$. For example, for the linear optimization problem

$$\min x_1 + x_2 \quad \text{s.t.} \quad x \geq 0$$

one may put $f(x) := x_1 + x_2$, $g_1(x) := -x_1$ and $g_2(x) := -x_2$.

2.2 The C^1-Characterization of Convexity

Before we deal with the C^1-characterization of convexity in Sect. 2.2.2, we first introduce the necessary notation and some results for multidimensional first derivatives in Sect. 2.2.1.

2.2.1 Multidimensional First Derivatives

For a nonempty open set $U \subseteq \mathbb{R}^n$ and a function $f : U \to \mathbb{R}$, we denote by $\partial_{x_i} f(\bar{x})$ the partial derivative of f with respect to x_i at the point $\bar{x} \in U$ (provided the derivative exists). For example, for $U = \mathbb{R}^2$ and $f(x) = x_1^2 + x_2$ we have

$\partial_{x_1} f(x) = 2x_1$ and $\partial_{x_2} f(x) = 1$. The evaluation at the point $\bar{x} = (1, -1)^\mathsf{T}$ yields $\partial_{x_1} f(\bar{x}) = 2$ and $\partial_{x_2} f(\bar{x}) = 1$.

The *first derivative* of f at $\bar{x}$ is considered to be the *row vector*

$$Df(\bar{x}) := (\partial_{x_1} f(\bar{x}), \ldots, \partial_{x_n} f(\bar{x})).$$

The *column vector* $\nabla f(\bar{x}) := (Df(\bar{x}))^\mathsf{T}$ is called the *gradient* of f at $\bar{x}$. In the case $n = 1$, we have $Df(\bar{x}) = \nabla f(\bar{x}) = f'(\bar{x})$. For example, for $f(x) = x_1^2 + x_2$ the first derivative and gradient at $\bar{x} = (1, -1)^\mathsf{T}$ are

$$Df(\bar{x}) = (2, 1) \quad \text{and} \quad \nabla f(\bar{x}) = \binom{2}{1},$$

respectively.

The function f is called *continuously differentiable* on U if ∇f exists on U and is a continuous function of x. We then write briefly $f \in C^1(U, \mathbb{R})$. For a not necessarily open set $X \subseteq \mathbb{R}^n$, the requirement $f \in C^1(X, \mathbb{R})$ means that there is some open superset $U \supseteq X$ with $f \in C^1(U, \mathbb{R})$.

For a *vector-valued* function

$$f : \mathbb{R}^n \to \mathbb{R}^m, \quad x \mapsto \begin{pmatrix} f_1(x) \\ \vdots \\ f_m(x) \end{pmatrix}$$

we define the first derivative at $\bar{x}$ as

$$Df(\bar{x}) := \begin{pmatrix} Df_1(\bar{x}) \\ \vdots \\ Df_m(\bar{x}) \end{pmatrix}.$$

This is an (m, n)-matrix, the *Jacobian matrix* or *derivative matrix* of f at $\bar{x}$. For example, for the function

$$f(x) = \begin{pmatrix} x_1^2 + x_2 \\ x_1 - x_2^2 \\ x_1 x_2 \end{pmatrix}$$

we obtain the Jacobian matrix

$$Df(x) = \begin{pmatrix} 2x_1 & 1 \\ 1 & -2x_2 \\ x_2 & x_1 \end{pmatrix}.$$

2.2 The C^1-Characterization of Convexity

Its evaluation at $\bar{x} = (1, -1)^\mathsf{T}$ is

$$Df(\bar{x}) = \begin{pmatrix} 2 & 1 \\ 1 & 2 \\ -1 & 1 \end{pmatrix}.$$

An important rule for differentiable functions is the *chain rule*, the proof of which can be found, for example, in [17].

Theorem 2.2.1 (Chain Rule) *Let $g : \mathbb{R}^n \to \mathbb{R}^m$ be differentiable at $\bar{x} \in \mathbb{R}^n$ and $f : \mathbb{R}^m \to \mathbb{R}^k$ be differentiable at $g(\bar{x}) \in \mathbb{R}^m$. Then $f \circ g : \mathbb{R}^n \to \mathbb{R}^k$ is differentiable at $\bar{x}$ with*

$$D(f \circ g)(\bar{x}) = Df(g(\bar{x})) \cdot Dg(\bar{x}).$$

A major reason for defining the Jacobian matrix of a function as above is that the chain rule can then be formulated completely analogously to the one-dimensional case ($n = m = k = 1$), although the appearing product is a matrix product.

The following theorem is proven, for example, in [14, 17] and illustrated in Fig. 2.4. As common in the literature, here we use the alternative notation $\langle a, b \rangle$ for the inner product $a^\mathsf{T} b$ of two vectors $a, b \in \mathbb{R}^n$.

Theorem 2.2.2 (Linearization by Taylor's Theorem in $\mathbb{R}^n$) *For a nonempty, open and convex set $U \subseteq \mathbb{R}^n$, let the function $f : U \to \mathbb{R}$ be differentiable at $x \in U$. Then for all $y \in U$*

$$f(y) = f(x) + \langle \nabla f(x), y - x \rangle + o(\|y - x\|)$$

holds, where $o(\|y - x\|)$ denotes an expression of the form $\omega(y)\|y - x\|$ with a function ω being continuous at x and satisfying $\omega(x) = 0$.

Fig. 2.4 Linear approximation of f around x for $n = 1$

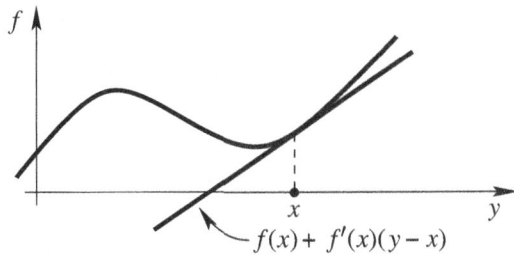

The 'qualitative' error term in Theorem 2.2.2 can be specified somewhat more explicitly with the help of the *Lagrange remainder term*:

$$o(\|y - x\|) = \langle \nabla f(\xi), y - x \rangle - \langle \nabla f(x), y - x \rangle,$$

where ξ is a point strictly in the connecting line segment between x and y, not known more explicitly. The resulting statement

$$f(y) = f(x) + \langle \nabla f(\xi), y - x \rangle$$

is also known as the *mean value theorem*.

2.2.2 C^1-Characterization

The following theorem plays a central role in the investigation of smooth convex functions. For *nonsmooth* convex functions, its statement is used to define the convex *subdifferential* as a substitute for the first derivative [34].

Theorem 2.2.3 (C^1-Characterization of Convexity) *On a convex set $X \subseteq \mathbb{R}^n$, a function $f \in C^1(X, \mathbb{R})$ is convex if and only if*

$$\forall x, y \in X: \quad f(y) \geq f(x) + \langle \nabla f(x), y - x \rangle$$

holds.

Proof Let f be convex on X, and let U be a convex open superset of X, on which f is continuously differentiable. Then according to Theorem 2.2.2 (with $z := (1 - \lambda)x + \lambda y$ in the role of y) for all $x, y \in X$ and $\lambda \in (0, 1)$

$$(1 - \lambda) f(x) + \lambda f(y) \geq f((1 - \lambda)x + \lambda y) = f(z)$$
$$= f(x) + \langle \nabla f(x), z - x \rangle + o(\|z - x\|)$$
$$= f(x) + \lambda \langle \nabla f(x), y - x \rangle + o(\lambda \|y - x\|)$$

holds, where $o(\|z - x\|)$ denotes an expression of the form $\omega(z)\|z - x\|$ with a function ω being continuous at x and satisfying $\omega(x) = 0$. After rearrangement and division by λ, this results in

$$f(y) \geq f(x) + \langle \nabla f(x), y - x \rangle + \omega((1 - \lambda)x + \lambda y)\|y - x\|.$$

2.2 The C^1-Characterization of Convexity

The limit $\lambda \to 0$ yields the desired inequality for all $x, y \in X$ due to the continuity of ω at x and $\omega(x) = 0$.

On the other hand, let $x, y \in X$, $\lambda \in (0, 1)$ and again $z := (1 - \lambda)x + \lambda y$. Then the two inequalities

$$f(x) \geq f(z) + \langle \nabla f(z), x - z \rangle,$$
$$f(y) \geq f(z) + \langle \nabla f(z), y - z \rangle$$

hold. Their convex combination yields

$$(1 - \lambda)f(x) + \lambda f(y) \geq f(z) + \langle \nabla f(z), \underbrace{(1 - \lambda)(x - z) + \lambda(y - z)}_{= (1-\lambda)x + \lambda y - z = 0} \rangle$$
$$= f(z) = f((1 - \lambda)x + \lambda y)$$

and thus the convexity of f on X. □

The C^1-characterization states that a C^1-function is convex on X if and only if its graph lies *above* each of its tangent spaces (Fig. 2.5).

A thorough review of the proof of Theorem 2.2.3 shows that the *continuity* of the first derivative of f is nowhere required. It is therefore also sufficient to assume a differentiable function f on X. However, most differentiable functions encountered in applications are also continuously differentiable, so the C^1-requirement is not overly restrictive and has become the established standard.

C^1-characterizations of *strict* and *strong* convexity are also known (e.g., [19, Theorem 4.1.1]).

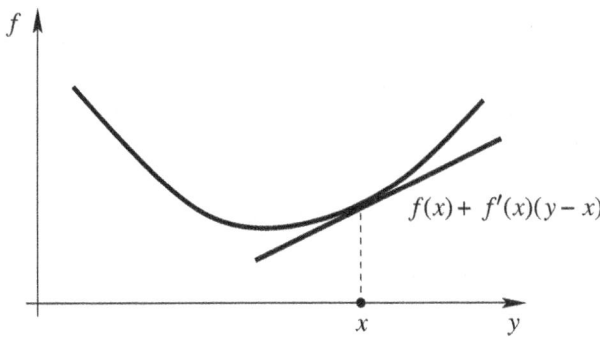

Fig. 2.5 C^1-characterization of convexity for $n = 1$

2.3 Solvability of Convex Optimization Problems

For the solvability of P, convexity alone is not a significant advantage: Both $f_1(x) = (x-5)^2$ and $f_2(x) = e^x$ are even strictly convex on $\mathbb{R}$, but only f_1 has a global minimal point on $\mathbb{R}$. In the following, we will see that the decisive advantage of f_1 over f_2 is its *strong* convexity.

Exercise 2.3.1 For a convex set $X \subseteq \mathbb{R}^n$ let $f : X \to \mathbb{R}$ be strongly convex. Show that f is then also strictly convex on X.

Lemma 2.3.2 *For a closed and convex set $X \subseteq \mathbb{R}^n$ let $f : X \to \mathbb{R}$ be strongly convex. Then f is also*

(a) coercive on X and
(b) continuous on the interior of X.

Proof We give the basic proof idea for statement a under the additional assumption of continuous differentiability of f on X.

For $X = \emptyset$ there is nothing to show. Otherwise, in the following we assume without loss of generality that $0 \in X$ holds (otherwise replace the variable x by $y = x - \bar{x}$ with some $\bar{x} \in X$ and argue that the assertion is independent of this shift). Due to the strong convexity of f on X, for some $c > 0$ the function $F(x) := f(x) - \frac{c}{2}\|x\|_2^2$ is convex on X. Rearranging the latter definition yields

$$f(x) = F(x) + \frac{c}{2}\|x\|_2^2.$$

To conclude $\lim_{\|x\|_2 \to \infty} f(x) = +\infty$ from this, $F(x)$ must not tend 'too quickly' towards $-\infty$ for $\|x\|_2 \to +\infty$. To verify this, we may use the C^1-characterization of convexity from Theorem 2.2.3, since f and therefore also F are continuously differentiable on X:

$$\forall x \in X: \quad F(x) \geq F(0) + \langle \nabla F(0), x \rangle \geq f(0) - \|\nabla f(0)\|_2 \cdot \|x\|_2,$$

where the second estimate follows from the Cauchy-Schwarz inequality and the definition of F. This means that F can 'drop at most at linear speed' towards $-\infty$ for $\|x\|_2 \to +\infty$.

Overall, we obtain

$$\forall x \in X: \quad f(x) = F(x) + \frac{c}{2}\|x\|_2^2 \geq f(0) - \|\nabla f(0)\|_2 \cdot \|x\|_2 + \frac{c}{2}\|x\|_2^2,$$

from which the coercivity of f on X follows.

2.4 Optimality Conditions for Unconstrained Convex Problems

If f is not continuously differentiable, the proof proceeds completely analogously by considering an element of the subdifferential of F instead of the C^1-characterization of F, the introduction of which we omit in the context of this textbook and instead refer to [3, 19, 34].

There it is also shown that every function that is convex on X is continuous on the interior of X. According to Exercise 2.3.1, the strongly convex function f is also strictly convex and thus in particular convex on X, so that assertion b follows. □

Theorem 2.3.3 *Let the optimization problem P be convex. Then the following statements hold:*

(a) The set of minimal points S is convex.
(b) If f is strictly convex on M, then S possesses at most one element.
(c) Let M be nonempty and closed, and f be strongly convex and continuous on M. Then S possesses exactly one element (i.e., P is uniquely solvable).

Proof In the case $S = \emptyset$ (e.g. $f(x) = e^x$, $M = \mathbb{R}$) statements a and b are trivially true. So let $S \neq \emptyset$. Then there exists some $\bar{x} \in S$, and with $v = f(\bar{x})$ it holds $S = \text{lev}^v_{\leq}(f, M)$. With Exercise 2.1.7 this implies statement a.

To prove statement b, we assume that there exists a point $\tilde{x} \in S \setminus \{\bar{x}\}$. Then it holds $f(\bar{x}) = f(\tilde{x}) = v$ and according to statement a also $f(\frac{1}{2}\bar{x} + \frac{1}{2}\tilde{x}) = v$. The strict convexity of f on M thus generates the contradiction

$$v = f\left(\frac{1}{2}\bar{x} + \frac{1}{2}\tilde{x}\right) < \frac{1}{2}f(\bar{x}) + \frac{1}{2}f(\tilde{x}) = v.$$

For the proof of statement c, Exercise 2.3.1 and statement b imply that S contains *at most* one element. According to Lemma 2.3.2a, f is coercive on M so that, by Corollary 1.2.33, S also contains *at least* one element.

□

We remark that, according to Lemma 2.3.2b, the continuity requirement for f in Theorem 2.3.3c is unnecessary if the set M coincides with its interior (under the conditions of Theorem 2.3.3c thus for $M = \mathbb{R}^n$).

2.4 Optimality Conditions for Unconstrained Convex Problems

In this section we will consider the unconstrained problem

$$P: \quad \min f(x).$$

More generally, problems with an *open* feasible set M are also referred to as unconstrained (roughly speaking, because no 'boundary effects' can occur in these problems). In fact, the optimality conditions discussed in the following can be easily transferred to this case which, for the sake of clarity, we will not explicitly state.

Definition 2.4.1 (Critical Point) A point $\bar{x} \in \mathbb{R}^n$ is called *critical* (or *stationary*) for $f \in C^1(\mathbb{R}^n, \mathbb{R})$, if $\nabla f(\bar{x}) = 0$ holds.

The following fundamental theorem holds without a convexity assumption and is proven for example in [37].

Theorem 2.4.2 (Fermat's Rule—First-Order Necessary Optimality Condition) *Let the point $\bar{x} \in \mathbb{R}^n$ by locally minimal for $f \in C^1(\mathbb{R}^n, \mathbb{R})$. Then $\bar{x}$ is a critical point of f.*

For example, $\bar{x} = (0,0)^\intercal$ is both a local minimal point and a critical point of the function $f_1(x) = x_1^2 + x_2^2$. However, $\bar{x} = (0,0)^\intercal$ is *no* local minimal point, but *nevertheless* a critical point of the function $f_2(x) = -x_1^2 - x_2^2$ and also of $f_3(x) = x_1^2 - x_2^2$. Consequently, not every critical point is necessarily a local minimal point of a C^1-function f. Since, in addition, not every *local* minimal point of a general function f is necessarily a *global* minimal point, Theorem 2.4.2 provides the set diagram in Fig. 2.6.

On the other hand, in the following two examples these relationships are sufficiently powerful to determine global minimal points.

Fig. 2.6 Necessary optimality condition in the unconstrained case

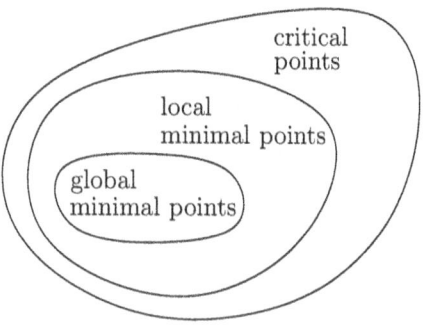

2.4 Optimality Conditions for Unconstrained Convex Problems

Example 2.4.3 (Center of a Point Cloud—Sequel 4) The function f from Example 1.1.5 satisfies

$$f(z) = \left\| \begin{pmatrix} \|z - x^1\|_2 \\ \vdots \\ \|z - x^m\|_2 \end{pmatrix} \right\|_2 = \sqrt{\sum_{i=1}^{m} \|z - x^i\|_2^2}$$

$$= \sqrt{\sum_{i=1}^{m} \underbrace{(z - x^i)^\mathsf{T}(z - x^i)}_{z^\mathsf{T} z - 2z^\mathsf{T} x^i + (x^i)^\mathsf{T} x^i}} = \sqrt{mz^\mathsf{T} z - 2z^\mathsf{T} \sum_{i=1}^{m} x^i + \sum_{i=1}^{m} \|x^i\|_2^2},$$

so that f is not differentiable, and Theorem 2.4.2 cannot be applied. However, by Exercise 1.3.5 with $\psi(y) = y^2$ and $Y = \{y \in \mathbb{R} | y \geq 0\}$, f has the same minimal points as the continuously differentiable function

$$f^2(z) = mz^\mathsf{T} z - 2z^\mathsf{T} \sum_{i=1}^{m} x^i + \sum_{i=1}^{m} \|x^i\|_2^2.$$

Critical points of this function are exactly the solutions of the equation

$$0 = \nabla(f^2(z)) = 2mz - 2\sum_{i=1}^{m} x^i,$$

so f^2 possesses the unique *critical* point

$$\bar{z} = \frac{1}{m} \sum_{i=1}^{m} x^i.$$

In fact, $\bar{z}$ is also the unique *global minimal* point of f^2 as well as of f, as can be seen from the following argument: By Example 1.2.34, a global minimal point $\tilde{z}$ of f and thus also of f^2 on $\mathbb{R}^n$ exists in the first place. Due to Fermat's rule (Theorem 2.4.2), $\tilde{z}$ must be a critical point of f^2. The *only* critical point of f^2 is the just calculated $\bar{z}$, so that $\tilde{z} = \bar{z}$ must hold. Thus, $\bar{z}$ is the unique global minimal point of f on $\mathbb{R}^n$.

In Sect. 2.5, we will show that f^2 is a convex function on $\mathbb{R}^n$, which will allow us to prove the global minimality of $\bar{z}$ in an alternative manner, *without* first struggling with the solvability of the underlying optimization problem.

Example 2.4.4 (Maximum Likelihood Estimator—Sequel 3) The objective function $f(\lambda) = \lambda \bar{x} - \log(\lambda)$ with $\bar{x} > 0$ of the problem P_{ML} from Example 1.2.39 fulfills $f'(\lambda) = \bar{x} - 1/\lambda$ and possesses the unique critical point $\bar{\lambda} = 1/\bar{x}$.

As in Example 2.4.3, it can be argued that $\bar{\lambda} = 1/\bar{x}$ is the unique global minimal point of P_{ML} and thus the sought-after maximum-likelihood estimator of the exponential distribution: According to Example 1.2.44, a global minimal point $\widetilde{\lambda}$ of P_{ML} exists in the first place. Since optimization problems with *open* feasible sets can be treated like *unconstrained* optimization problems, $\widetilde{\lambda}$ must be a critical point of f according to Fermat's rule. The *only* critical point of f is the just calculated $\bar{\lambda}$, from which the assertion follows.

In Sect. 2.5, we will also see that f is convex on the feasible set $(0, +\infty)$ of P_{ML}, which will again allow us to prove the global minimality of $\bar{\lambda}$ alternatively, *without* first showing the solvability of P_{ML}.

The alternative proofs suggested in the preceding examples are based on the fact that the relationship between critical points and global minimal points becomes considerably clearer if $f \in C^1(\mathbb{R}^n, \mathbb{R})$ is additionally *convex*.

> **Theorem 2.4.5 (First-Order Sufficient Optimality Condition)** *Let the function $f \in C^1(\mathbb{R}^n, \mathbb{R})$ be convex. Then every critical point of f is a global minimal point of f.*

Proof Let the point x be critical for f, i.e., it holds $\nabla f(x) = 0$. With Theorem 2.2.3 it follows

$$\forall\, y \in \mathbb{R}^n: \quad f(y) \geq f(x) + \underbrace{\langle \nabla f(x), y - x \rangle}_{=0} = f(x),$$

thus the assertion. □

Figure 2.7 shows the relation between critical points and global minimal points proven in Theorem 2.4.5 in a set diagram. Overall, we obtain the following important result.

Fig. 2.7 Sufficient optimality condition in the unconstrained case

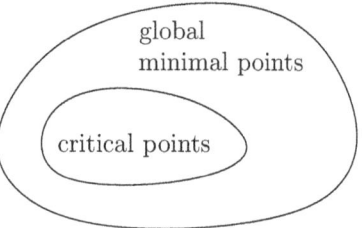

2.5 The C^2-Characterization of Convexity

Corollary 2.4.6 (First-Order Characterization of Global Minimal Points)
Let the function $f \in C^1(\mathbb{R}^n, \mathbb{R})$ be convex. Then the global minimal points are exactly the critical points of f.

Proof Theorems 2.4.2 and 2.4.5. □

To determine global minimal points of unconstrained convex C^1-problems, it is therefore not only sufficient to search for local minimal points (as already seen in Theorem 2.1.6), but even only for critical points. The global minimization problem is thus reduced to solving the equation $\nabla f(x) = 0$.

In particular, one also obtains a statement about solvability: The function f has a global minimal point on $\mathbb{R}^n$ if and only if f has a critical point. For example, the convex C^1-function $f(x) = e^x$ satisfies the inequality $f'(x) = e^x > 0$ for all $x \in \mathbb{R}$, so it does not have a critical point. It is clear, therefore, that f cannot possess a global minimal point on $\mathbb{R}$.

2.5 The C^2-Characterization of Convexity

Before we discuss C^2-characterizations of the different types of convexity in Sect. 2.5.2, we first introduce the necessary notation and some results for multi-dimensional second derivatives in Sect. 2.5.1.

2.5.1 The Multidimensional Second Derivative

For a nonempty open set $U \subseteq \mathbb{R}^n$ and $f : U \to \mathbb{R}$, the *second derivative* is defined as the first derivative of the gradient (if it exists):

$$D^2 f(x) := D\nabla f(x) = \begin{pmatrix} \partial_{x_1}\partial_{x_1} f(x) & \cdots & \partial_{x_n}\partial_{x_1} f(x) \\ \vdots & & \vdots \\ \partial_{x_1}\partial_{x_n} f(x) & \cdots & \partial_{x_n}\partial_{x_n} f(x) \end{pmatrix}.$$

For example, the function $f(x) = x_1^2 + x_2$ satisfies

$$D^2 f(x) = D\begin{pmatrix} 2x_1 \\ 1 \end{pmatrix} = \begin{pmatrix} 2 & 0 \\ 0 & 0 \end{pmatrix}.$$

The matrix $D^2 f(\bar{x})$ is called the *Hessian matrix* of f at $\bar{x}$ and is always an (n,n)-matrix. If all entries of $D^2 f$ are continuous functions of x, f is called *twice continuously differentiable* on U, shortly $f \in C^2(U, \mathbb{R})$. In this case, $D^2 f(x)$ is

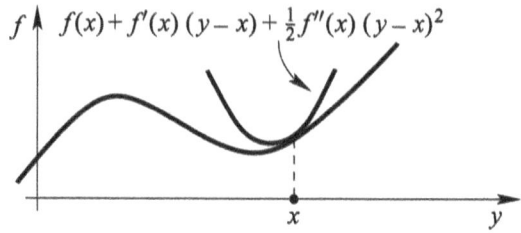

Fig. 2.8 Quadratic approximation of f around x for $n = 1$

even symmetric for $x \in U$ (according to Schwarz's theorem [17]). The requirement $f \in C^2(X, \mathbb{R})$ with an arbitrary set $X \subseteq \mathbb{R}^n$ means again that there is some open superset $U \supseteq X$ with $f \in C^2(U, \mathbb{R})$. For $n = 1$ we have $D^2 f(\bar{x}) = f''(\bar{x})$.

Also the next theorem is shown, for example, in [14, 17]. Figure 2.8 illustrates the result.

Theorem 2.5.1 (Quadratic Approximation by Taylor's Theorem in $\mathbb{R}^n$)
For a nonempty, open and convex set $U \subseteq \mathbb{R}^n$, let the function $f : U \to \mathbb{R}$ be twice differentiable at $x \in U$. Then for all $y \in U$

$$f(y) = f(x) + \langle \nabla f(x), y - x \rangle + \tfrac{1}{2}(y-x)^\mathsf{T} D^2 f(x)(y-x) + o(\|y-x\|^2)$$

holds. The qualitative error term can be specified using the Lagrange remainder term*:*

$$o(\|y - x\|^2) = \tfrac{1}{2}(y - x)^\mathsf{T} D^2 f(\xi)(y - x) - \tfrac{1}{2}(y - x)^\mathsf{T} D^2 f(x)(y - x),$$

where ξ is some point strictly in the connecting line segment between x and y, which is not known more explicitly.

In the context of optimization problems, decisive properties of the Hessian matrix of a function are given by the following terms. An (n, n)-matrix A is called *positive semidefinite* (shorthand: $A \succeq 0$), if

$$\forall d \in \mathbb{R}^n : \quad d^\mathsf{T} A d \geq 0$$

applies, and *positive definite* ($A \succ 0$), if this inequality is even strict for all $d \neq 0$. Positive (semi)definiteness of a matrix is often difficult to verify by definition. In linear algebra, for *symmetric* matrices A, it is fortunately shown [9, 22] that $A \succeq 0$ ($\succ 0$) is fulfilled if and only if $\lambda \geq 0$ (> 0) applies for all *eigenvalues* λ of A (see, for example, the appendix of [30] for an introduction to eigenvalues). This is considerably easier to verify. For $n = 1$, A collapses to a scalar, and the equivalence of $A \succeq 0$ ($\succ 0$) with $A \geq 0$ (> 0) is then easily seen.

2.5 The C^2-Characterization of Convexity

With this concept we can formulate the following optimality condition, which applies without convexity assumptions and is proven by Theorem 2.5.1 (e.g. [37]).

Theorem 2.5.2 (Second-Order Necessary Optimality Condition) *Let the point $\bar{x}$ be a local minimal point of $f : \mathbb{R}^n \to \mathbb{R}$, and let f be twice differentiable at $\bar{x}$. Then $\nabla f(\bar{x}) = 0$ and $D^2 f(\bar{x}) \succeq 0$ hold.*

2.5.2 C^2-Characterizations

This section provides useful conditions to check convexity, strict convexity, and strong convexity of functions, provided they are twice continuously differentiable.

Theorem 2.5.3 (C^2-Characterization of Convexity) *On a convex set $X \subseteq \mathbb{R}^n$, let the function $f \in C^2(X, \mathbb{R})$ be given.*

(a) If
$$\forall x \in X : \quad D^2 f(x) \succeq 0$$
holds, then f is convex on X.
(b) If, in addition, X is open, then the converse of statement a also holds.

Proof To prove statement a, we first choose some convex open superset U of X, on which f is twice continuously differentiable. According to Theorem 2.5.1, it then holds in particular for all $x, y \in X \subseteq U$ with some ξ on the line segment between x and y

$$f(y) = f(x) + \langle \nabla f(x), y - x \rangle + \frac{1}{2}(y - x)^\mathsf{T} D^2 f(\xi)(y - x)$$
$$\geq f(x) + \langle \nabla f(x), y - x \rangle,$$

where we have used $D^2 f(\xi) \succeq 0$. This is allowed because, due to the convexity of X, with x and y the entire line segment between x and y is contained in X, so in particular the point ξ. According to Theorem 2.2.3, f is therefore convex on X.

To prove statement b, choose some $\bar{x} \in X$. With f, the function $F(x) := f(x) - \langle \nabla f(\bar{x}), x - \bar{x} \rangle$ is also convex on X, where $\bar{x}$ is to be interpreted as a fixed parameter. From $\nabla F(\bar{x}) = \nabla f(\bar{x}) - \nabla f(\bar{x}) = 0$ follows that $\bar{x}$ is a critical point of F. Since X is open, as mentioned above the same statements apply to the constrained minimization of F over X as for the *unconstrained* minimization of F. In particular, $\bar{x}$ is not only a critical point, but according to Corollary 2.4.6 even a global minimal point of F on X. Theorem 2.5.2 therefore guarantees

$$0 \preceq D^2 F(\bar{x}) = D^2 f(\bar{x}).$$

□

The openness requirement for X in Theorem 2.5.3b is not just for proof-technical reasons, as can be seen from the C^2-function $f(x) = x_1^2 - x_2^2$, which does *not* have a positive semidefinite Hessian matrix anywhere, but is nevertheless convex on the set $X = \mathbb{R} \times \{0\}$. In this example the set X is not open.

The openness requirement for X in Theorem 2.5.3b can be weakened to the *full-dimensionality* of X, which is essentially the requirement that X possesses interior points [33]. A further weakening is not possible.

Example 2.5.4 The function $f(x) = (x-5)^2$ satisfies $f''(x) = 2 \geq 0$ for all $x \in \mathbb{R}$ and is therefore convex on $\mathbb{R}$.

Example 2.5.5 The function $f(x) = e^x$ satisfies $f''(x) = e^x \geq 0$ for all $x \in \mathbb{R}$ and is therefore convex on $\mathbb{R}$.

Example 2.5.6 (Center of a Point Cloud—Sequel 5) For the gradient of the function f^2 from Example 2.4.3 we have already derived the representation

$$\nabla(f^2(z)) = 2mz - 2\sum_{i=1}^{m} x^i$$

from which

$$D^2(f^2(z)) = 2mI \quad \text{(with the } (n, n)\text{-identity matrix } I\text{)}$$

follows. Thus, $D^2 f^2(z)$ has the n-fold eigenvalue $2m \geq 0$. This implies the convexity of f^2 on $\mathbb{R}^n$.

According to Corollary 2.4.6, the critical points of f^2 are exactly the global minimal points of f^2 and therefore also of f, and the unique critical point of f^2 is the arithmetic mean $\bar{z}$ of the point cloud calculated in Example 2.4.3. This alternatively demonstrates the global minimality of $\bar{z}$, without first proving the solvability of the optimization problem.

Example 2.5.7 (Maximum Likelihood Estimator—Sequel 4) The function $f(\lambda) = \lambda \bar{x} - \log(\lambda)$ with $\bar{x} > 0$ from Example 2.4.4 satisfies $f'(\lambda) = \bar{x} - 1/\lambda$ and $f''(\lambda) = 1/\lambda^2 \geq 0$ for all $\lambda \in (0, +\infty)$. Therefore, it is convex on $(0, +\infty)$. The convexity of f corresponds to the concavity of the log-likelihood function ℓ. Indeed, the likelihood function L itself is *not* concave.

2.5 The C^2-Characterization of Convexity

The logarithmization of L therefore has another useful effect that we had not initially intended: It reveals that the maximum likelihood problem ML is hidden convex in the sense that it can be equivalently reformulated as a convex minimization problem.

Again, the critical point $\bar{\lambda} = 1/\bar{x}$ of f calculated in Example 2.4.4 is, according to Corollary 2.4.6, a global minimal point of P_{ML} and thus the sought-after maximum likelihood estimator of the exponential distribution. Also in this argument (alternatively to that in Example 2.4.4) the consideration of the solvability of P_{ML} is unnecessary.

Exercise 2.5.8 (Cluster Analysis—Sequel 3) Show that the objective function

$$f(z^1, \ldots, z^q) = \sum_{i=1}^{m} \min_{\ell=1,\ldots,q} \|z^\ell - x^i\| = \left\| \begin{pmatrix} \|z^{\ell(1)} - x^1\| \\ \vdots \\ \|z^{\ell(m)} - x^m\| \end{pmatrix} \right\|_1$$

from the problem P_1 of cluster analysis (Example 1.1.8) with $q \geq 2$ is not convex on $\mathbb{R}^{nq}$ (however, if the indices $\ell(i)$, $i = 1, \ldots, m$, are known in advance, then with a different argument than using Theorem 2.5.3, it can be shown that f is indeed convex).

Exercise 2.5.9 (Sufficient Condition for Strict Convexity) On a convex set $X \subseteq \mathbb{R}^n$ let the function $f \in C^2(X, \mathbb{R})$ be given, and let

$$\forall x \in X: \quad D^2 f(x) \succ 0$$

be true. Show that f is then strictly convex on X.

The reversal of the statement in Exercise 2.5.9 is even false on open sets X, as the example of the strictly convex function $f(x) = x^4$ on $X = \mathbb{R}$ with $f''(0) = 0$ shows. However, a C^2-characterization of strict convexity does exist and is provided in [33].

In the following, let $\lambda_{\min}(A)$ denote the smallest eigenvalue of a symmetric matrix A. In particular, it holds $D^2 f(x) \succeq 0$ ($\succ 0$) exactly for $\lambda_{\min}(D^2 f(x)) \geq 0$ (> 0).

Theorem 2.5.10 (C^2-Characterization of Strong Convexity) *On a convex set $X \subseteq \mathbb{R}^n$ let the function $f \in C^2(X, \mathbb{R})$ be given.*

(a) *If with a constant $c > 0$*

$$\forall x \in X: \quad \lambda_{\min}(D^2 f(x)) \geq c$$

holds, then f is strongly convex on X.
(b) *If, in addition, X is open, then the reversal of statement a also holds.*

Proof To show statement a, we construct the function $F(x) = f(x) - \frac{c}{2}\|x\|_2^2$ using the constant c and show its convexity on X. By definition, f is then strongly convex on X.

Due to $\|x\|_2^2 = x^\mathsf{T} x$, with f also the function F is twice continuously differentiable on X, so that its convexity can be proven using Theorem 2.5.3a. In fact, for all $x \in X$

$$D^2 F(x) = D^2 f(x) - cI,$$

holds, with the (n,n)-identity matrix I. Every eigenvalue λ of $D^2 f(x)$ satisfies the equation $\det(D^2 f(x) - \lambda I) = 0$ (for a motivation see the appendix of [30]), so we obtain

$$0 = \det(D^2 f(x) - \lambda I) = \det((D^2 f(x) - cI) - (\lambda - c)I)$$
$$= \det(D^2 F(x) - (\lambda - c)I).$$

This means that every eigenvalue of $D^2 F(x)$ can be written in the form $\lambda - c$ with an eigenvalue λ of $D^2 f(x)$. In particular, $\lambda_{\min}(D^2 F(x)) = \lambda_{\min}(D^2 f(x)) - c$ holds. Since the last expression is nonnegative for all $x \in X$ according to the assumption, the convexity of F on X follows.

The proof of statement b is left as an exercise for the reader. □

The requirement of an open set X in Theorem 2.5.10b can again be weakened to the *full-dimensionality* of X [33], but no further.

Example 2.5.11 The function $f(x) = (x - 5)^2$ satisfies $f''(x) = 2 > 0$ for all $x \in \mathbb{R}$ and is therefore not only strictly, but even strongly convex on $\mathbb{R}$.

Example 2.5.12 The function $f(x) = e^x$ satisfies $f''(x) = e^x > 0$ for all $x \in \mathbb{R}$ and is therefore strictly convex on $\mathbb{R}$. However, in view of $\lim_{x \to -\infty} f''(x) = 0$ it is not strongly convex on $\mathbb{R}$.

Example 2.5.13 (Center of a Point Cloud—Sequel 6) For the Hessian matrix of the square of the function f from Example 1.1.5, we derived in Example 2.5.6 the representation $D^2 f^2(z) = 2mI$ and thus the n-fold eigenvalue $2m$ for each $z \in \mathbb{R}^n$. In particular, then $\lambda_{\min}(D^2 f^2(z)) = 2m > 0$ holds for all $z \in \mathbb{R}^n$, from which the strict and even strong convexity of f^2 on $\mathbb{R}^n$ follow.

Even without the previous calculation of the global minimal point, it is now clear from Theorem 2.3.3c that f^2 (and thus also f) has exactly one global minimal point.

Example 2.5.14 (Maximum Likelihood Estimator—Sequel 5) The function $f(\lambda) = \lambda \bar{x} - \log(\lambda)$ with $\bar{x} > 0$ from Example 1.2.39 satisfies $f''(\lambda) = 1/\lambda^2 > 0$ for all $\lambda \in (0, +\infty)$, but $\lim_{\lambda \to +\infty} f''(\lambda) = 0$. It is therefore strictly, but not strongly convex on $(0, +\infty)$.

2.6 The Monotonicity Characterization of Convexity

According to Example 1.2.41, f is nevertheless coercive. This shows that the condition from Lemma 2.3.2 is only sufficient, but not necessary for coercivity.

2.6 The Monotonicity Characterization of Convexity

For $n = 1$, an open interval $X \subseteq \mathbb{R}$ and a function $g \in C^1(X, \mathbb{R})$ it is known from calculus (e.g. [18]) that g is monotonically increasing on X if and only if $g'(x) \geq 0$ holds for all $x \in X$. Hence, according to Theorem 2.5.3, for $n = 1$, an open interval $X \subseteq \mathbb{R}$ and $f \in C^2(X, \mathbb{R})$, f is convex on X if and only if the first derivative f' is monotonically increasing on X.

In the following, we will see that such a statement also holds without the C^2-assumption for f and more generally for $n \geq 1$. However, we first need to define what we mean by monotonicity of the function $\nabla f : X \to \mathbb{R}^n$ for a set $X \subseteq \mathbb{R}^n$. To do this, we note that for $n = 1$ monotonicity (in the sense of monotone increase) of $g : X \to \mathbb{R}$ is equivalent to the validity of the inequality $(g(y) - g(x))(y - x) \geq 0$ for all $x, y \in X$. This motivates the following definition.

Definition 2.6.1 (Monotone Operator) For a nonempty convex set $X \subseteq \mathbb{R}^n$ the operator $g : X \to \mathbb{R}^n$ is called *monotone* on X, if

$$\forall x, y \in X : \quad \langle g(y) - g(x), y - x \rangle \geq 0$$

holds.

Theorem 2.6.2 (Monotonicity Characterization of Convexity) *On a convex set $X \subseteq \mathbb{R}^n$, a function $f \in C^1(X, \mathbb{R})$ is convex if and only if ∇f is monotone on X.*

Proof Let the function $f \in C^1(X, \mathbb{R})$ be convex on X. According to Theorem 2.2.3, for all $x, y \in X$

$$f(y) \geq f(x) + \langle \nabla f(x), y - x \rangle \quad \text{and} \quad f(x) \geq f(y) + \langle \nabla f(y), x - y \rangle$$

hold. Adding these two inequalities implies the monotonicity of ∇f on X.

On the other hand, let ∇f be monotone on X. We choose arbitrary $x, y \in X$, set $d := y - x$ as well as $x(t) := x + td$ for $t \in \mathbb{R}$ and consider the one-dimensional restriction

$$\varphi_d : [0, 1] \to \mathbb{R}, \ t \mapsto f(x(t))$$

of f at x in direction d [37] (which is continuously differentiable on an open superset of the interval $[0, 1]$). The chain rule yields $\varphi'_d(t) = \langle \nabla f(x(t)), d \rangle$ for each $t \in [0, 1]$. Since $x(t) - x(0) = td$ holds, from the monotonicity of ∇f we obtain for each $t \in (0, 1]$

$$\varphi'_d(t) - \varphi'_d(0) = \langle \nabla f(x(t)) - \nabla f(x(0)), d \rangle$$
$$= \frac{1}{t} \langle \nabla f(x(t)) - \nabla f(x(0)), x(t) - x(0) \rangle \geq 0.$$

Therefore, the mean value theorem yields the existence of some $t \in (0, 1)$ with

$$f(y) - f(x) = \varphi_d(1) - \varphi_d(0) = \varphi'_d(t) \geq \varphi'_d(0) = \langle \nabla f(x), y - x \rangle,$$

so that the convexity of f on X follows from Theorem 2.2.3. □

2.7 Optimality Conditions for Constrained Convex Problems

In this section, we attempt to transfer the simple characterization of global minimal points by the critical point equation $\nabla f(x) = 0$ from the unconstrained case considered in Sect. 2.4 to constrained optimization problems of the form

$$P: \quad \min f(x) \ \text{s.t.} \ g_i(x) \leq 0, \ i \in I, \quad h_j(x) = 0, \ j \in J.$$

For this purpose, in a first step, we neither assume convexity nor differentiability of the functions f, g_i, $i \in I$, and h_j, $j \in J$. We explicitly write the index sets as $I = \{1, \ldots, p\}$ and $J = \{1, \ldots, q\}$ with $p, q \in \mathbb{N}_0$, where $p = 0$ and $q = 0$ correspond to the cases $I = \emptyset$ and $J = \emptyset$, respectively. Since under certain regularity conditions [37] each equality constraint reduces the dimension of the solution space by one, and since we want to model an at least one-dimensional feasible set

$$M = \{x \in \mathbb{R}^n | \ g_i(x) \leq 0, \ i \in I, \ h_j(x) = 0, \ j \in J\},$$

we also assume $q < n$.

Even for the case $n = 1$, Fig. 1.5 shows that at minimal points $\bar{x}$ of constrained problems, $\nabla f(\bar{x}) = 0$ does not necessarily hold. One-dimensional problems like in Fig. 1.5 can often be solved by considering the boundary points of the feasible set as candidates for minimal points in addition to the critical points of the objective function, because the set of boundary points in one dimension is typically finite. However, this algorithmic approach is not transferable to higher-dimensional

2.7 Optimality Conditions for Constrained Convex Problems

problems, where the boundary of the feasible set usually contains infinitely many points itself. Hence we need to find a way to filter out the 'critical' boundary points.

The necessary optimality conditions for constrained optimization problems are closely linked to duality results, which we first derive in Sect. 2.7.1. In Sect. 2.7.2, they lead to the central *sufficient* optimality conditions for constrained differentiable optimization problems, namely the Karush-Kuhn-Tucker conditions. Compared to the critical point condition $\nabla f(x) = 0$ from the unconstrained case, they are algorithmically more challenging, mainly because of the occurrence of complementarity conditions, which are discussed in Sect. 2.7.3. With the geometric interpretation of the Karush-Kuhn-Tucker conditions in Sect. 2.7.4 we motivate the formulation of constraint qualifications in Sect. 2.7.5, which are employed in the characterization of global minimal points by the Karush-Kuhn-Tucker conditions.

We only state as a result that local minimal points of differentiable constrained optimization problems under constraint qualifications also *necessarily* satisfy the Karush-Kuhn-Tucker conditions and refer to [37] for the rather extensive derivation, as this result is not attributable to global, but to nonlinear (local) optimization.

2.7.1 Lagrange and Wolfe Duality

The basis for the statements on Lagrange and Wolfe duality is the following aggregation of all defining functions of an optimization problem P.

Definition 2.7.1 (Lagrange Function) The function
$$L(x, \lambda, \mu) = f(x) + \sum_{i \in I} \lambda_i g_i(x) + \sum_{j \in J} \mu_j h_j(x)$$
(with $\lambda = (\lambda_1, \ldots, \lambda_p)^\mathsf{T}$ and $\mu = (\mu_1, \ldots, \mu_q)^\mathsf{T}$) is called *Lagrange function* of the optimization problem P.

It is important to note that the Lagrange function L depends not only on the decision variable x, but also on the coefficient vectors λ and μ. The function L thus maps from $\mathbb{R}^n \times \mathbb{R}^p \times \mathbb{R}^q$ to $\mathbb{R}$. In the following, we investigate what happens when L is *maximized* over the variables (λ, μ) for some fixed x.

For this, let us first assume $p = 0$. Then $L(x, \mu) = f(x) + \sum_{j \in J} \mu_j h_j(x)$ and

$$\varphi(x) := \sup_{\mu \in \mathbb{R}^q} L(x, \mu) = \begin{cases} f(x), & \text{if } \forall j \in J : h_j(x) = 0 \text{ (i.e., if } x \in M) \\ +\infty, & \text{if } x \notin M \end{cases}$$

Fig. 2.9 Example for
$\varphi(x) := \sup_{\mu \in \mathbb{R}^q} L(x, \mu)$

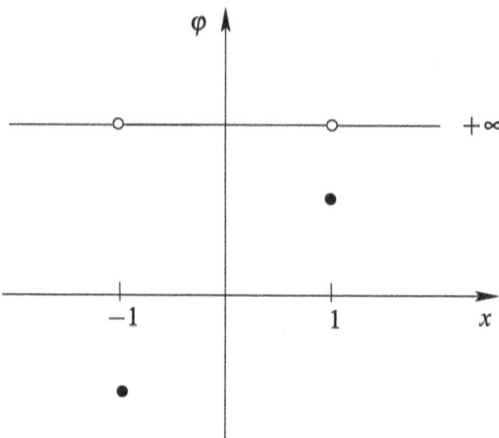

hold. For example, $J = \{1\}$, $h_1(x) = x^2 - 1$ and $f(x) = x$ lead to the function φ sketched in Fig. 2.9 (where for once we allow $q = n$ in this example).

Instead of solving P, one can formally also minimize the *unconstrained* function $\varphi(x)$ ('formally', because φ does not map to $\mathbb{R}$, but to the extended real numbers $\overline{\mathbb{R}}$). In particular, the infimum v of f on M satisfies

$$v = \inf_{x \in M} f(x) = \inf_{x \in \mathbb{R}^n} \varphi(x) = \inf_{x \in \mathbb{R}^n} \sup_{\mu \in \mathbb{R}^q} L(x, \mu).$$

Next, let $q = 0$. Then $L(x, \lambda) = f(x) + \sum_{i \in I} \lambda_i g_i(x)$,

$$\varphi(x) := \sup_{\lambda \geq 0} L(x, \lambda) = \begin{cases} f(x), & \text{if } x \in M \\ +\infty, & \text{if } x \notin M \end{cases}$$

and

$$v = \inf_{x \in M} f(x) = \inf_{x \in \mathbb{R}^n} \varphi(x) = \inf_{x \in \mathbb{R}^n} \sup_{\lambda \geq 0} L(x, \lambda)$$

hold. Observe that without the nonnegativity condition $\lambda \geq 0$ this result would not be true. As an example, $I = \{1\}$, $g_1(x) = x^2 - 1$ and $f(x) = x$ result in the function φ sketched in Fig. 2.10.

Analogously, for any $p, q \in \mathbb{N}_0$

$$\varphi(x) := \sup_{\lambda \geq 0, \, \mu \in \mathbb{R}^q} L(x, \lambda, \mu) = \begin{cases} f(x), & \text{if } x \in M \\ +\infty, & \text{if } x \notin M \end{cases}$$

2.7 Optimality Conditions for Constrained Convex Problems

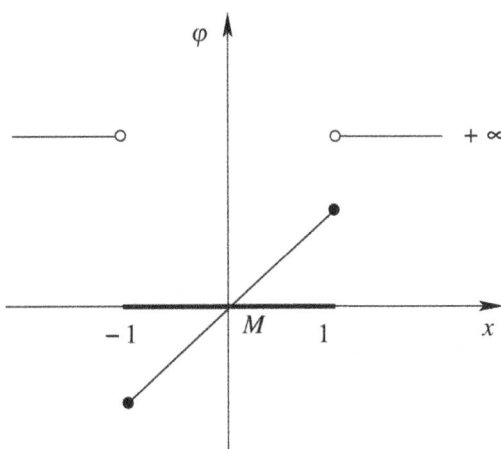

Fig. 2.10 Example for
$\varphi(x) := \sup_{\lambda \geq 0} L(x, \lambda)$

and

$$v = \inf_{x \in M} f(x) = \inf_{x \in \mathbb{R}^n} \varphi(x) = \inf_{x \in \mathbb{R}^n} \sup_{\lambda \geq 0,\, \mu \in \mathbb{R}^q} L(x, \lambda, \mu)$$

hold.

The *central question* of duality theory (and, e.g., also of noncooperative game theory) is whether in the above formula inf and sup can be interchanged, i.e. whether also the formula

$$v = \inf_{x \in M} f(x) = \sup_{\lambda \geq 0,\, \mu \in \mathbb{R}^q} \inf_{x \in \mathbb{R}^n} L(x, \lambda, \mu)$$

is true. If, namely, the inner unconstrained infimum of L over x (given some (λ, μ)) is easily computable, one obtains v as the maximal value of a constrained problem with simple restrictions, which allows important conclusions. The following example unfortunately shows that this interchange is not always possible (compare this with the statement of Exercise 1.3.3d).

Example 2.7.2 For $n = 1$, $p = 1$, $q = 0$, $f(x) = x^2$ and $g(x) = 1 - x^4 \leq 0$ one calculates $v = 1$. On the other hand, we have

$$L(x, \lambda) = x^2 + \lambda(1 - x^4)$$

and thus

$$\inf_{x \in \mathbb{R}} L(x, \lambda) = \begin{cases} -\infty, & \lambda > 0 \\ 0, & \lambda = 0 \end{cases}$$

as well as

$$\sup_{\lambda \geq 0} \inf_{x \in \mathbb{R}} L(x, \lambda) = 0 < 1 = v.$$

Duality theory provides conditions under which such examples are excluded, i.e. equality holds. For this, we consider the expression

$$\sup_{\lambda \geq 0,\, \mu \in \mathbb{R}^q} \inf_{x \in \mathbb{R}^n} L(x, \lambda, \mu)$$

as the supremum of the optimization problem

$$LD: \quad \max_{\lambda, \mu} \psi(\lambda, \mu) \quad \text{s.t.} \quad \lambda \geq 0$$

with the objective function

$$\psi(\lambda, \mu) := \inf_{x \in \mathbb{R}^n} L(x, \lambda, \mu).$$

Definition 2.7.3 (Lagrange Dual) The problem *LD* is called *Lagrange dual* of *P*. We denote its feasible set with

$$M_{LD} := \left\{ (\lambda, \mu) \in \mathbb{R}^p \times \mathbb{R}^q \mid \lambda \geq 0 \right\}$$

and its maximal value with

$$v_{LD} := \sup_{(\lambda, \mu) \in M_{LD}} \psi(\lambda, \mu).$$

Analogously, we denote *P* as the *primal* problem with minimal value $v_P := v$ and feasible set $M_P := M$.

In this notation the central question of duality theory is when the equality

$$v_{LD} = v_P$$

is fulfilled. *Without further assumptions*, the following theorem is correct (compare this to the statement of Exercise 1.3.3c).

2.7 Optimality Conditions for Constrained Convex Problems

Theorem 2.7.4 (Weak Duality Theorem of Lagrange Duality) *The inequality $v_{LD} \leq v_P$ holds.*

Proof For all $\bar{x} \in \mathbb{R}^n$, $\bar{\lambda} \geq 0$ and $\bar{\mu} \in \mathbb{R}^q$ it holds

$$\inf_{x \in \mathbb{R}^n} L(x, \bar{\lambda}, \bar{\mu}) \leq L(\bar{x}, \bar{\lambda}, \bar{\mu}) \leq \sup_{\lambda \geq 0, \mu \in \mathbb{R}^q} L(\bar{x}, \lambda, \mu).$$

The reformulation of these infinitely many inequalities by supremum or infimum formation (Exercise 1.2.6) leads to the claimed relationship

$$\underbrace{\sup_{\bar{\lambda} \geq 0, \bar{\mu} \in \mathbb{R}^q} \inf_{x \in \mathbb{R}^n} L(x, \bar{\lambda}, \bar{\mu})}_{v_{LD}} \leq \underbrace{\inf_{\bar{x} \in \mathbb{R}^n} \sup_{\lambda \geq 0, \mu \in \mathbb{R}^q} L(\bar{x}, \lambda, \mu)}_{v_P}.$$

□

According to Theorem 2.7.4 it holds

$$v_P - v_{LD} \geq 0.$$

The value $v_P - v_{LD}$ is called *duality gap*. In Example 2.7.2 the duality gap is $v_P - v_{LD} = 1$. In this terminology, the central question of duality theory is when the duality gap vanishes.

In order to use duality computationally, the dual problem LD must first be made manageable, because in general it is unclear how the values of its objective function $\psi(\lambda, \mu) = \inf_{x \in \mathbb{R}^n} L(x, \lambda, \mu)$ can be calculated algorithmically. The key to this is the (so far not assumed) convexity of the problem P.

In the following, as in Example 2.1.12, let the functions f and g_i, $i \in I$, be convex on $\mathbb{R}^n$ and h_j, $j \in J$, linear, so that P is a convexly described optimization problem (in particular, we assume that its feasible set M is not only convex, but also convexly *described* in the sense of Definition 2.1.10). Additionally, let the functions f and g_i, $i \in I$, be continuously differentiable. We will somewhat loosely refer to this situation as P being 'convexly described and C^1'.

According to Exercise 2.1.3, for every fixed $\lambda \geq 0$ and $\mu \in \mathbb{R}^q$, the Lagrange function

$$L(x, \lambda, \mu) = f(x) + \sum_{i \in I} \lambda_i g_i(x) + \sum_{j \in J} \mu_j h_j(x)$$

is then convex on $\mathbb{R}^n$ as well as C^1 in the variable x. Consequently, for all $\lambda \geq 0$ and $\mu \in \mathbb{R}^q$ the value $\psi(\lambda, \mu) = \inf_{x \in \mathbb{R}^n} L(x, \lambda, \mu)$ is the infimum of an *unconstrained convex C^1-problem*. Therefore, if we can find a critical point of the function

$L(x, \lambda, \mu)$, i.e., a solution x^* of the system of equations

$$\nabla_x L(x, \lambda, \mu) = 0,$$

then, according to Corollary 2.4.6, x^* is also a global minimal point of $L(x, \lambda, \mu)$ over $x \in \mathbb{R}^n$. This implies

$$\psi(\lambda, \mu) = \inf_{x \in \mathbb{R}^n} L(x, \lambda, \mu) = \min_{x \in \mathbb{R}^n} L(x, \lambda, \mu) = L(x^*, \lambda, \mu).$$

In particular, the infimum in the definition of $\psi(\lambda, \mu)$ is then attained as the minimal value.

This motivates the formulation of a dual problem that differs from the Lagrange dual.

Definition 2.7.5 (Wolfe Dual) Let the problem P be convexly described and C^1. Then

$$D: \max_{x,\lambda,\mu} L(x, \lambda, \mu) \quad \text{s.t.} \quad \nabla_x L(x, \lambda, \mu) = 0, \; \lambda \geq 0$$

is called the *Wolfe dual* of P. We denote its feasible set with

$$M_D = \left\{ (x, \lambda, \mu) \in \mathbb{R}^n \times \mathbb{R}^p \times \mathbb{R}^q \,\middle|\, \nabla_x L(x, \lambda, \mu) = 0, \; \lambda \geq 0 \right\},$$

and its maximal value with

$$v_D = \sup_{(x,\lambda,\mu) \in M_D} L(x, \lambda, \mu).$$

Compared to the Lagrange dual, the Wolfe dual is often an algorithmically better manageable optimization problem, although it is not necessarily convex itself. We denote it as 'D' instead of 'WD' to distinguish it from the Lagrange dual LD, as the Lagrange dual will not play a role in the rest of this textbook, while we will derive all occurring duality statements with the help of the Wolfe dual D.

We will henceforth call a point $x \in M_P := M$ *primal feasible* and $(x, \lambda, \mu) \in M_D$ *dual feasible*. Weak duality between the primal problem P and its Wolfe dual D could be inferred from Theorem 2.7.4 with some formal arguments, but for completeness, we provide a direct proof.

2.7 Optimality Conditions for Constrained Convex Problems

Theorem 2.7.6 (Weak Duality Theorem of Wolfe Duality) *For every convexly described C^1-problem P, $v_D \leq v_P$ holds.*

Proof In the case $M_D = \emptyset$ we have $v_D = -\infty$, so the inequality is formally correct. Similarly, it is fulfilled in the case $M_P = \emptyset$, due to $v_P = +\infty$. Otherwise, we choose some arbitrary point $x \in M_P$ and some arbitrary point $(y, \lambda, \mu) \in M_D$. This implies in particular $\lambda \geq 0$, so that $L(z, \lambda, \mu)$ is a convex C^1-function in $z \in \mathbb{R}^n$. Moreover, $\nabla_x L(y, \lambda, \mu) = 0$ holds, which is why y is a global minimal point of $L(z, \lambda, \mu)$ over all $z \in \mathbb{R}^n$. This yields

$$L(y, \lambda, \mu) = \min_{z \in \mathbb{R}^n} L(z, \lambda, \mu).$$

From $\lambda \geq 0$, $g_i(x) \leq 0$, $i \in I$, and $h_j(x) = 0$, $j \in J$, it follows

$$f(x) \geq f(x) + \sum_{i \in I} \lambda_i g_i(x) + \sum_{j \in J} \mu_j h_j(x) = L(x, \lambda, \mu)$$
$$\geq \min_{z \in \mathbb{R}^n} L(z, \lambda, \mu) = L(y, \lambda, \mu).$$

This finally implies

$$v_P = \inf_{x \in M_P} f(x) \geq \sup_{(y, \lambda, \mu) \in M_D} L(y, \lambda, \mu) = v_D,$$

thus the assertion. $\square$

As seen in the proof, remarkably the inequality $v_D \leq v_P$ is meaningful even for $M_D = \emptyset$ and/or $M_P = \emptyset$, due to the formal definitions of suprema and infima over empty sets. If either the primal or the dual optimization problem is unbounded, the application of weak duality to these definitions formally even implies that the other problem must be inconsistent. However, since one cannot 'calculate' with extended real numbers in this way, an alternative proof for this result must be sought.

Exercise 2.7.7 Show without recourse to extended real values of infima and suprema that the unboundedness of the primal or of the Wolfe dual optimization problem implies the inconsistency of the Wolfe dual or the primal optimization problem, respectively.

The following theorem indicates how to generate explicitly calculable bounds on minimal values using weak duality.

Theorem 2.7.8 *Let a convexly described C^1-problem P be given.*

(a) Let the point $\tilde{x}$ be primal feasible. Then $f(\tilde{x})$ is an upper bound for the global minimal value of P:

$$v_P \leq f(\tilde{x}).$$

(b) Let the point $(\bar{x}, \bar{\lambda}, \bar{\mu})$ be dual feasible. Then $L(\bar{x}, \bar{\lambda}, \bar{\mu})$ is a lower bound for the global minimal value of P:

$$v_P \geq L(\bar{x}, \bar{\lambda}, \bar{\mu}).$$

Proof For the proof of statement a no recourse to duality theory is necessary:

$$f(\tilde{x}) \overset{\tilde{x} \in M_P}{\geq} \inf_{x \in M} f(x) = v_P.$$

Furthermore, from weak duality follows

$$L(\bar{x}, \bar{\lambda}, \bar{\mu}) \overset{(\bar{x},\bar{\lambda},\bar{\mu}) \in M_D}{\leq} \sup_{(x,\lambda,\mu) \in M_D} L(x, \lambda, \mu) = v_D \leq v_P,$$

thus statement b. □

The following example illustrates that Theorem 2.7.8, which is based solely on weak duality, may already allow far-reaching results.

Example 2.7.9 (Distance from a Hyperplane) We are looking for the distance $\text{dist}(z, H)$ from $z \in \mathbb{R}^n$ to the hyperplane $H = \{x \in \mathbb{R}^n | a^\mathsf{T} x = b\}$ with $a \in \mathbb{R}^n \setminus \{0\}$ and $b \in \mathbb{R}$ (as in Example 1.1.2, only here we are not primarily interested in an optimal *point*, but as in Example 1.2.10 in the optimal *value* of the projection problem). Figure 2.11 illustrates the problem for $n = 2$, $a = (1, 2)^\mathsf{T}$, $b = 2$ and $z = 0$. In the following, we will see which statements weak duality allows for the general problem

$$P: \min_{x \in \mathbb{R}^n} \|x - z\|_2 \quad \text{s.t.} \quad a^\mathsf{T} x = b$$

and illustrate this with the above special example

$$\min_{x \in \mathbb{R}^2} \|x\|_2 \quad \text{s.t.} \quad x_1 + 2x_2 = 2$$

(the corresponding specialized results are given in parentheses).

2.7 Optimality Conditions for Constrained Convex Problems

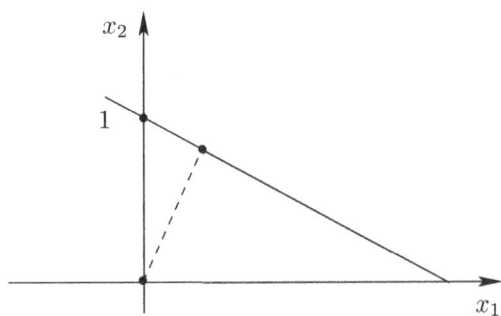

Fig. 2.11 Distance from a hyperplane in the case $n = 2$

First, by squaring the objective function, according to Exercise 1.3.5 the nondifferentiable problem P is equivalent to

$$Q: \quad \min_x (x-z)^\mathsf{T}(x-z) \quad \text{s.t.} \quad a^\mathsf{T} x = b.$$

In Q, the objective function $f(x) := x^\mathsf{T} x - 2z^\mathsf{T} x + z^\mathsf{T} z$ is convex and C^1 on $\mathbb{R}^n$, and the equality constraint function $h(x) = a^\mathsf{T} x - b$ is linear. So Q is convexly described and C^1 and thus suitable for the application of Wolfe duality.

The Lagrange function of Q is

$$L(x, \mu) = x^\mathsf{T} x - 2z^\mathsf{T} x + z^\mathsf{T} z + \mu(a^\mathsf{T} x - b),$$

from which

$$\nabla_x L(x, \mu) = 2x - 2z + \mu a$$

follows. With this, the Wolfe dual to Q can be set up:

$$D: \quad \max_{x \in \mathbb{R}^n,\ \mu \in \mathbb{R}} x^\mathsf{T} x - 2z^\mathsf{T} x + z^\mathsf{T} z + \mu(a^\mathsf{T} x - b) \quad \text{s.t.} \quad 2(x-z) + \mu a = 0$$

$$\left(D: \quad \max_{x \in \mathbb{R}^2,\ \mu \in \mathbb{R}} x_1^2 + x_2^2 + \mu(x_1 + 2x_2 - 2) \quad \text{s.t.} \quad 2x + \mu \begin{pmatrix} 1 \\ 2 \end{pmatrix} = 0 \right).$$

In view of $M_P = H$, primal feasibility of x just means $a^\mathsf{T} x = b$. Dual feasibility of (x, μ) holds exactly for $2(x-z) + \mu a = 0$.

Upper Bound for v_P

Since for example $\bar{x} = (b/a^\mathsf{T} a)\, a$ is a primal feasible point, Theorem 2.7.8a provides the estimate

$$v_Q \le f\left(b\,\frac{a}{a^\mathsf{T} a}\right) = b^2\,\frac{a^\mathsf{T} a}{(a^\mathsf{T} a)^2} - 2z^\mathsf{T}\left(b\,\frac{a}{a^\mathsf{T} a}\right) + z^\mathsf{T} z$$

$$= \frac{b^2}{a^\mathsf{T} a} - \frac{2b}{a^\mathsf{T} a} z^\mathsf{T} a + z^\mathsf{T} z.$$

It follows

$$v_P = \sqrt{v_Q} \le \sqrt{\frac{b^2}{a^\mathsf{T} a} - \frac{2b}{a^\mathsf{T} a} z^\mathsf{T} a + z^\mathsf{T} z}$$

$$\left(v_Q \le \frac{4}{5},\quad v_P \le \frac{2}{\sqrt{5}}\right).$$

(In the specific example, another choice of a primal feasible point is $\bar{x} = (0, 1)^\mathsf{T}$, from which $v_Q \le f(\bar{x}) = 1$ and $v_P \le 1$ follow. Compared to the just calculated value, this is however a worse upper bound.)

Lower Bound for v_P

In the condition $2(x - z) + \mu a = 0$ for dual feasibility, z and a are given parameters, while pairs (x, μ) are sought that fulfill this condition. By isolating x one obtains

$$x = z - \frac{\mu}{2} a,$$

thereby providing for each arbitrarily chosen $\bar{\mu} \in \mathbb{R}$ the matching point $\bar{x} = z - (\bar{\mu}/2)\, a$ for dual feasibility. It follows:

$$\forall\, \bar{\mu} \in \mathbb{R}:\quad (\bar{x}, \bar{\mu}) = \left(z - \frac{\bar{\mu}}{2} a,\ \bar{\mu}\right) \in M_D.$$

According to Theorem 2.7.8b, for all $\bar{\mu} \in \mathbb{R}$ we obtain

$v_Q \ge L(\bar{x}, \bar{\mu})$

$$= L\left(z - \frac{\bar{\mu}}{2} a,\ \bar{\mu}\right)$$

$$= \left(z - \frac{\bar{\mu}}{2} a\right)^\mathsf{T}\left(z - \frac{\bar{\mu}}{2} a\right) - 2z^\mathsf{T}\left(z - \frac{\bar{\mu}}{2} a\right) + z^\mathsf{T} z + \bar{\mu}\left(a^\mathsf{T}\left(z - \frac{\bar{\mu}}{2} a\right) - b\right)$$

$$= \bar{\mu}(z^\mathsf{T} a - b) - \frac{\bar{\mu}^2}{4} a^\mathsf{T} a$$

2.7 Optimality Conditions for Constrained Convex Problems

$$\left(v_Q \geq -2\bar\mu - \frac{5}{4}\bar\mu^2, \quad \text{e.g.} \quad \bar\mu = 0 \Rightarrow v_Q \geq 0 \quad \text{(which is clear anyway)},\right.$$

$$\bar\mu = 1 \Rightarrow v_Q \geq -\frac{13}{4} \quad \text{(which is even clearer)},$$

$$\left.\bar\mu = -1 \Rightarrow v_Q \geq \frac{3}{4}\right).$$

In view of $v_Q \geq 0$ this implies for all $\bar\mu \in \mathbb{R}$

$$v_Q \geq \max\left\{0,\ \bar\mu(z^\mathsf{T} a - b) - \frac{\bar\mu^2}{4} a^\mathsf{T} a\right\}$$

and thus

$$v_P = \sqrt{v_Q} \geq \sqrt{\max\left\{0,\ \bar\mu(z^\mathsf{T} a - b) - \frac{\bar\mu^2}{4} a^\mathsf{T} a\right\}}$$

$$\left(v_P \geq \sqrt{\{\max\{0,\ -2\bar\mu - \frac{5}{4}\bar\mu^2\}}\}, \text{ e.g. } \bar\mu = -1: \underbrace{\frac{\sqrt{3}}{2}}_{\approx 0.86} \leq v_P \overset{\text{see above}}{\leq} \underbrace{\frac{2}{\sqrt{5}}}_{\approx 0.89}\right).$$

As the *best* possible lower bound for v_Q one obtains with this approach the supremum of the bounds over all choices of $\bar\mu \in \mathbb{R}$,

$$\sup_{\bar\mu \in \mathbb{R}} \bar\mu(z^\mathsf{T} a - b) - \frac{\bar\mu^2}{4} a^\mathsf{T} a,$$

which is exactly the maximal value of the dual problem D. For the calculation of v_D we set

$$c(\bar\mu) := \bar\mu(z^\mathsf{T} a - b) - \frac{\bar\mu^2}{4} a^\mathsf{T} a$$

and obtain

$$c'(\bar\mu) = a^\mathsf{T} z - b - \frac{\bar\mu}{2} a^\mathsf{T} a,$$
$$c''(\bar\mu) = -\frac{a^\mathsf{T} a}{2} \overset{a \neq 0}{<} 0.$$

Consequently, the function c is concave. According to Corollary 2.4.6 its global maximal points are exactly the solutions of

$$0 = c'(\bar\mu) = a^\mathsf{T} z - b - \frac{\bar\mu}{2} a^\mathsf{T} a,$$

i.e., the unique maximal point of c is

$$\bar{\mu}^\star = 2\frac{a^\mathsf{T} z - b}{a^\mathsf{T} a}$$

with maximal value

$$v_D = c(\bar{\mu}^\star) = 2\frac{(a^\mathsf{T} z - b)^2}{a^\mathsf{T} a} - \frac{(a^\mathsf{T} z - b)^2}{a^\mathsf{T} a} = \frac{(a^\mathsf{T} z - b)^2}{a^\mathsf{T} a}.$$

From this, as the best lower bound for v_P achievable by Wolfe duality follows

$$v_P = \sqrt{v_Q} \geq \sqrt{v_D} = \frac{|a^\mathsf{T} z - b|}{\|a\|_2}$$

$$\left(v_P \geq \frac{|-2|}{\sqrt{5}} = \frac{2}{\sqrt{5}} \overset{\text{see above}}{\geq} v_P \right).$$

In the specific example, these considerations result in $v_P = \frac{2}{\sqrt{5}}$, because *coincidentally* we had previously also found the best upper bound. In the general case, however, weak duality only provides an estimate for the distance from z to H:

$$v_P \geq \frac{|a^\mathsf{T} z - b|}{\|a\|_2}.$$

Next, an improvement of this estimate to an identity as well as a formula for the minimal *point* of P would be desirable.

To this end, the following lemma is helpful, provided its requirements can be met.

Lemma 2.7.10 *Let the optimization problem P be convexly described and C^1, and let $\bar{x} \in \mathbb{R}^n$, $\bar{\lambda} \in \mathbb{R}^p$, $\bar{\mu} \in \mathbb{R}^q$ be given with*

(a) *$\bar{x}$ is primal feasible,*
(b) *$(\bar{x}, \bar{\lambda}, \bar{\mu})$ is dual feasible,*
(c) *$f(\bar{x}) = L(\bar{x}, \bar{\lambda}, \bar{\mu})$.*

Then $\bar{x}$ is a global minimal point of P.

2.7 Optimality Conditions for Constrained Convex Problems

Proof It holds

$$f(\bar{x}) \overset{a}{\geq} \inf_{x \in M_P} f(x) = v_P \geq v_D = \sup_{(x,\lambda,\mu) \in M_D} L(x, \lambda, \mu) \overset{b}{\geq} L(\bar{x}, \bar{\lambda}, \bar{\mu})$$
$$\overset{c}{=} f(\bar{x}).$$

Therefore, every inequality in this chain must be satisfied with equality. This means in particular $f(\bar{x}) = v_P$ (and also that the duality gap vanishes). Thus, $\bar{x}$ is a global minimal point of P.

$\square$

Example 2.7.11 (Distance from a Hyperplane—Sequel 1) As seen above, $(\bar{x}, \bar{\mu})$ is dual feasible if one chooses $\bar{x} = z - (\bar{\mu}/2)\, a$ for any $\bar{\mu} \in \mathbb{R}$. Moreover, primal feasibility of $\bar{x}$ means that the equation $a^\mathsf{T} \bar{x} = b$ is satisfied. Thus, for condition a and b from Lemma 2.7.10 to hold simultaneously, $\bar{x}$ and $\bar{\mu}$ must satisfy the system of equations

$$\bar{x} = z - \frac{\bar{\mu}}{2} a,$$
$$a^\mathsf{T} \bar{x} = b$$

The solution is calculated to be

$$\mu^\star = 2 \frac{a^\mathsf{T} z - b}{a^\mathsf{T} a},$$
$$x^\star = z - \frac{a^\mathsf{T} z - b}{a^\mathsf{T} a} a.$$

Although $(x^\star, \mu^\star)$ is already uniquely determined (z, a and b are given), condition c from Lemma 2.7.10 must also hold, i.e.

$$f(x^\star) = L(x^\star, \mu^\star).$$

Fortunately, the already determined pair $(x^\star, \mu^\star)$ satisfies this equation in view of

$$L(x^\star, \mu^\star) = f(x^\star) + \mu^\star \underbrace{(a^\mathsf{T} x^\star - b)}_{=0} = f(x^\star).$$

In total, conditions a, b and c from Lemma 2.7.10 are fulfilled, so that

$$x^\star = z - \frac{a^\mathsf{T} z - b}{a^\mathsf{T} a} a$$

is a global minimal point of Q with minimal value

$$v_Q = \frac{(a^\mathsf{T} z - b)^2}{a^\mathsf{T} a}.$$

As seen, $x^\star$ is then also a global minimal point of P with minimal value

$$v_P = \frac{|a^\mathsf{T} z - b|}{\|a\|_2}.$$

The fact that, as in the above example, condition c from Lemma 2.7.10 is automatically fulfilled by the solutions from condition a and b is indeed *not* a coincidence, but this effect occurs in every problem *without inequality constraints*.

Corollary 2.7.12 *Let the optimization problem P be convexly described and C^1, $I = \emptyset$, and let $\bar{x} \in \mathbb{R}^n$, $\bar{\mu} \in \mathbb{R}^q$ be given with*

(a) $\bar{x}$ is primal feasible,
(b) $(\bar{x}, \bar{\mu})$ is dual feasible.

Then $\bar{x}$ is a global minimal point of P.

Proof It holds

$$M_P = \{x \in \mathbb{R}^n \mid h_j(x) = 0, \ j \in J\}$$

and thus

$$L(\bar{x}, \bar{\mu}) = f(\bar{x}) + \sum_{j \in J} \bar{\mu}_j h_j(\bar{x}) \stackrel{\bar{x} \in M_P}{=} f(\bar{x}).$$

In total, conditions a, b and c from Lemma 2.7.10 are fulfilled, from which the assertion follows. □

2.7.2 The Karush-Kuhn-Tucker Conditions

Since *inequalities* can also be strictly fulfilled, one cannot argue for them as in the proof of Corollary 2.7.12. However, condition c from Lemma 2.7.10 can still be written more simply, for which the following concept is introduced (where convexity initially plays *no* role).

2.7 Optimality Conditions for Constrained Convex Problems

Definition 2.7.13 (Karush-Kuhn-Tucker Point) For a C^1-problem P, a point $\bar{x} \in \mathbb{R}^n$ is called *Karush-Kuhn-Tucker point (KKT point)* with multipliers $\bar{\lambda}$ and $\bar{\mu}$, if the following system of equalities and inequalities is fulfilled:

$$\nabla_x L(\bar{x}, \bar{\lambda}, \bar{\mu}) = 0, \tag{2.1}$$

$$\bar{\lambda}_i g_i(\bar{x}) = 0, \ i \in I, \tag{2.2}$$

$$\bar{\lambda}_i \geq 0, \ i \in I, \tag{2.3}$$

$$g_i(\bar{x}) \leq 0, \ i \in I, \tag{2.4}$$

$$h_j(\bar{x}) = 0, \ j \in J. \tag{2.5}$$

Lemma 2.7.14 *The conditions a, b and c in Lemma 2.7.10 are satisfied for $(\bar{x}, \bar{\lambda}, \bar{\mu})$ if and only if $\bar{x}$ is a KKT point of P with multipliers $\bar{\lambda}, \bar{\mu}$.*

Proof We need to show the equivalence of conditions a, b and c from Lemma 2.7.10 with conditions (2.1) to (2.5) from Definition 2.7.13.

We first assume conditions a, b and c. From condition a, (2.4) and (2.5) follow, from condition b, (2.1) and (2.3) follow, and from condition c, we first obtain

$$\sum_{i \in I} \bar{\lambda}_i g_i(\bar{x}) + \sum_{j \in J} \bar{\mu}_j h_j(\bar{x}) = 0.$$

Since condition a implies $h_j(\bar{x}) = 0, \ j \in J$, this equation reduces to

$$\sum_{i \in I} \bar{\lambda}_i g_i(\bar{x}) = 0.$$

Furthermore, condition a implies $g_i(\bar{x}) \leq 0, \ i \in I$, and condition b $\bar{\lambda}_i \geq 0, \ i \in I$, so that each term in the above sum is nonpositive. For the sum to be zero, then none of the terms can be strictly negative, so we obtain

$$\bar{\lambda}_i g_i(\bar{x}) = 0, \ i \in I,$$

i.e., (2.2).

In the second step to prove the desired equivalence we assume conditions (2.1) to (2.5). From (2.4) and (2.5) condition a follows, from (2.1) and (2.3) condition b follows, and from (2.2) and (2.5) condition c follows. □

Fig. 2.12 Sufficient optimality condition in the constrained case

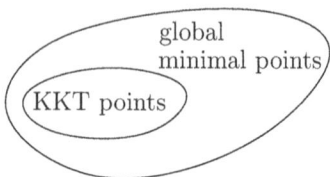

Theorem 2.7.15 (Sufficient Optimality Condition for P convex and C^1)
Let the optimization problem P be convexly described and C^1, and let $\bar{x}$ be a KKT point (with multipliers $\bar{\lambda}, \bar{\mu}$). Then $\bar{x}$ is a global minimal point of P.

Proof Lemmas 2.7.14 and 2.7.10. □

Observe that, by Lemmas 2.7.14 and 2.7.10c, in Theorem 2.7.15 the requirement that $\bar{x}$ is a KKT point in particular implies the requirement of a vanishing duality gap.

Example 2.7.16 (Distance from a Hyperplane—Sequel 2) The point $\bar{x}$ is a KKT point of Q with multiplier $\bar{\mu}$ if and only if the equations $2(\bar{x} - z) + \bar{\mu} a = 0$ and $a^\mathsf{T} \bar{x} - b = 0$ are satisfied. As above, this results in

$$x^\star = z - \frac{a^\mathsf{T} z - b}{a^\mathsf{T} a} a \quad \text{and} \quad \mu^\star = 2 \frac{a^\mathsf{T} z - b}{a^\mathsf{T} a}.$$

The representation of the result from Theorem 2.7.15 as a set diagram in Fig. 2.12 leads to an illustration analogous to Fig. 2.7.

This suggests that also the result illustrated in Fig. 2.6 may apply analogously, i.e., that all global minimal points are necessarily KKT points. The following example shows that this is unfortunately *not* the case.

Example 2.7.17 The problem

$$P: \quad \min x \quad \text{s.t.} \quad x^2 \leq 0$$

is convexly described and C^1 with $M_P = S = \{0\}$. However, $\bar{x} = 0$ is *not* a KKT point of P, because the Lagrange function of P is

$$L(x, \lambda) = x + \lambda x^2,$$

from which

$$\nabla_x L(x, \lambda) = 1 + 2\lambda x$$

2.7 Optimality Conditions for Constrained Convex Problems

follows. The equation of the KKT system

$$0 = \nabla_x L(\bar{x}, \lambda) = 1 + 2\lambda \cdot 0 = 1$$

is thus not solvable for any $\lambda \geq 0$.

This example shows that the expected analogue to Corollary 2.4.6, namely 'every global minimal point of P is a KKT point' does *not* apply for constrained problems.

In a negative formulation this also means: It is possible that a constrained problem has no KKT points, but is still solvable. Hence, even if one can prove that a problem has no KKT points, this does not mean that the problem is also unsolvable.

Exercise 2.7.18 Show that in Example 2.7.17 strong duality in the sense of the identity $v_P = v_D$ applies. Why can the condition $f(\bar{x}) = L(\bar{x}, \bar{\lambda})$ from Lemma 2.7.10c still not be fulfilled?

In the following, we investigate which additional conditions guarantee that a global minimal point is a KKT point.

2.7.3 Complementarity

As seen in the proof of Lemma 2.7.14, under conditions a and b from Lemma 2.7.10 (primal and dual feasibility) condition c (identity of primal and dual objective function value) is equivalent to

$$\bar{\lambda}_i \geq 0, \quad g_i(\bar{x}) \leq 0, \quad 0 = \bar{\lambda}_i g_i(\bar{x}), \quad i \in I.$$

For each $i \in I$ the condition

$$\bar{\lambda}_i \geq 0, \quad g_i(\bar{x}) \leq 0, \quad \bar{\lambda}_i g_i(\bar{x}) = 0$$

is called a *complementarity condition*.

Depending on whether the constraint $g_i(x) \leq 0$ is fulfilled at $\bar{x}$ with equality or with strict inequality, the complementarity condition has different consequences: In the case $g_i(\bar{x}) = 0$ the equality $\bar{\lambda}_i g_i(\bar{x}) = 0$ holds for any $\bar{\lambda}_i \geq 0$, i.e., the complementarity condition is automatically fulfilled. For $g_i(\bar{x}) < 0$, the condition $\bar{\lambda}_i g_i(\bar{x}) = 0$ implies $\bar{\lambda}_i = 0$. The term *complementarity* refers to the fact that at least one of the numbers $\bar{\lambda}_i$ and $g_i(\bar{x})$ vanishes.

The important distinction whether an inequality constraint is satisfied with equality or strict inequality motivates the following definition.

Fig. 2.13 Points with different active index sets

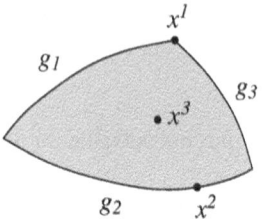

Definition 2.7.19 (Active Index Set) For $\bar{x} \in M$,

$$I_0(\bar{x}) = \{i \in I \mid g_i(\bar{x}) = 0\}$$

is called the *set of active indices* or briefly *active index set* of $\bar{x}$.

For example, in Fig. 2.13 we have

$$I_0(x^1) = \{1, 3\}, \quad I_0(x^2) = \{2\}, \quad I_0(x^3) = \emptyset.$$

2.7.4 Geometric Interpretation of the KKT Conditions

Since the complementarity conditions enforce for all $i \notin I_0(\bar{x})$ that $\bar{\lambda}_i$ vanishes, $\bar{x}$ is a KKT point with multipliers $\bar{\lambda}, \bar{\mu}$ if and only if the following system is satisfied:

$$\nabla f(\bar{x}) + \sum_{i \in I_0(\bar{x})} \bar{\lambda}_i \nabla g_i(\bar{x}) + \sum_{j \in J} \bar{\mu}_j \nabla h_j(\bar{x}) = 0,$$

$$g_i(\bar{x}) = 0, \ i \in I_0(\bar{x}),$$
$$g_i(\bar{x}) < 0, \ i \notin I_0(\bar{x}),$$
$$h_j(\bar{x}) = 0, \ j \in J,$$
$$\bar{\lambda}_i \geq 0, \ i \in I_0(\bar{x}),$$
$$\bar{\lambda}_i = 0, \ i \notin I_0(\bar{x}).$$

Because the multipliers to inactive inequalities $\bar{\lambda}_i, i \notin I_0(\bar{x})$, only appear in the last line of this system, they may as well be omitted. The resulting system is suitable for geometric considerations as well as for manual calculations in small problems. On the other hand, it is less suitable for algorithmic evaluations, because it explicitly depends on the a priori unknown active index set $I_0(\bar{x})$. It must therefore be treated

2.7 Optimality Conditions for Constrained Convex Problems

by case distinction according to the possible active index sets $I_0(\bar{x}) \subseteq I$. For a set I comprised of p elements, this results in 2^p cases.

To understand the geometric meaning of the Karush-Kuhn-Tucker conditions, we assume the knowledge of the fact that the gradients of a C^1-function stand perpendicular to its level sets and point in the direction of steepest ascent of the function values [37].

We first consider the case *without inequality constraints*, i.e. $I = \emptyset$. Then the KKT system reduces to

$$\nabla f(\bar{x}) + \sum_{j=1}^{q} \bar{\mu}_j \nabla h_j(\bar{x}) = 0,$$

$$h_j(\bar{x}) = 0, \ j \in J.$$

For $n = 2$ and $q = 1$, Fig. 2.14 shows level lines of $f(x) = x_1^2 + x_2^2$ and a feasible set described by a linear equality constraint $h(x) = 0$. Since the first equation in the KKT conditions states that ∇f and ∇h are linearly dependent, x^1 is not a KKT point, while x^2 is.

It is noteworthy that, although the geometry of the feasible set M does not change when h is replaced by $-h$, $\nabla(-h) = -\nabla h$ points in the opposite direction to ∇h. So even though changing the sign of h has no impact on whether $\bar{x}$ is, for example, a local minimal point, the KKT conditions change. However, this is compensated by the fact that the multipliers μ_j, $j \in J$, are not sign-constrained. The replacement of h_j by $-h_j$ can therefore simply be absorbed in the KKT conditions by replacing $\bar{\mu}_j$ with $-\bar{\mu}_j$.

Next we turn to the case *without equality constraints*, i.e. $J = \emptyset$. Then the KKT system reads

$$\nabla f(\bar{x}) + \sum_{i \in I_0(\bar{x})} \bar{\lambda}_i \nabla g_i(\bar{x}) = 0,$$

$$g_i(\bar{x}) = 0, \ i \in I_0(\bar{x}),$$

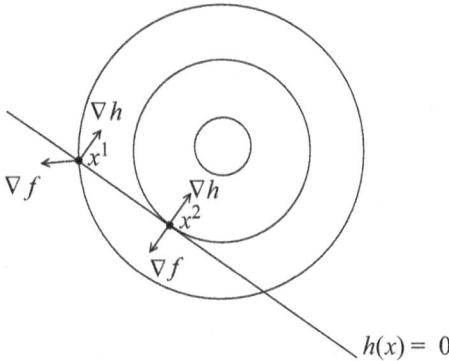

Fig. 2.14 KKT point for $I = \emptyset$

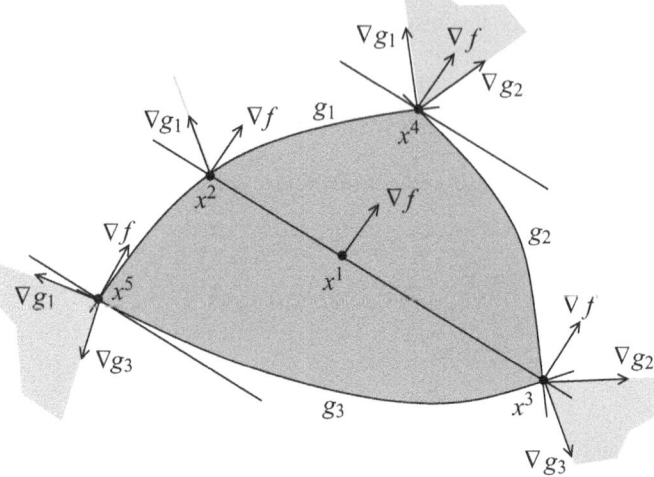

Fig. 2.15 KKT point for $J = \emptyset$

$$g_i(\bar{x}) < 0, \; i \notin I_0(\bar{x}),$$
$$\bar{\lambda}_i \geq 0, \; i \in I_0(\bar{x}).$$

This means in particular that the vector $-\nabla f(\bar{x})$ lies in the set

$$\text{cone}\left(\{\nabla g_i(\bar{x}), \; i \in I_0(\bar{x})\}\right) := \left\{ \sum_{i \in I_0(\bar{x})} \lambda_i \nabla g_i(\bar{x}) \,\middle|\, \lambda \geq 0 \right\}$$

i.e., the convex cone spanned by the vectors $\nabla g_i(\bar{x}), \; i \in I_0(\bar{x})$.

For $n = 2$ and $p = 3$, Fig. 2.15 shows a feasible set described by convex inequality constraints, level lines of a linear objective function f (e.g. $f(x) = x_1 + x_2$) and some convex cones spanned by gradients of active inequalities. By the above geometric interpretation, it is easy to see that among the points x^1 to x^5 in Fig. 2.15, only x^5 is a KKT point. Since P is a convex optimization problem, x^5 must even be a global minimal point.

With this geometric background one can also see that the KKT concept is not able to cover, for example, minimal points at cusps of the feasible set, as the one shown in Fig. 2.16. While, like in Example 2.7.17, $\bar{x}$ is then a global minimal point, it is *not* a KKT point.

2.7.5 Constraint Qualifications

These geometric considerations motivate the introduction of *constraint qualifications*, under which global minimal points are guaranteed to be KKT points.

2.7 Optimality Conditions for Constrained Convex Problems

Fig. 2.16 Feasible set with cusp

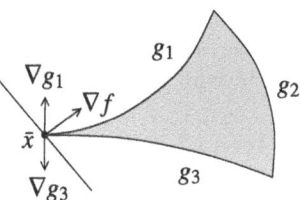

Since constraint qualifications are designed to ensure a nondegenerate structure of the feasible set [37], the objective function f of P does not play a role in their formulation.

Definition 2.7.20 (Constraint Qualifications)

(a) The *linear independence constraint qualification (LICQ)* holds at $\bar{x} \in M$ if the vectors $\nabla g_i(\bar{x}), i \in I_0(\bar{x}), \nabla h_j(\bar{x}), j \in J$, are linearly independent (i.e., the equation

$$\sum_{i \in I_0(\bar{x})} \lambda_i \nabla g_i(\bar{x}) + \sum_{j \in J} \mu_j \nabla h_j(\bar{x}) = 0$$

only has the solution $(\lambda, \mu) = (0, 0)$).

(b) The *Mangasarian-Fromovitz constraint qualifcation (MFCQ)* holds at $\bar{x} \in M$ if the vectors $\nabla h_j(\bar{x}), j \in J$, are linearly independent and the system

$$\langle \nabla g_i(\bar{x}), d \rangle < 0, \; i \in I_0(\bar{x}), \; \langle \nabla h_j(\bar{x}), d \rangle = 0, \; j \in J,$$

has a solution d (which, as shown in e.g. [37], is equivalent to the system

$$\sum_{i \in I_0(\bar{x})} \lambda_i \nabla g_i(\bar{x}) + \sum_{j \in J} \mu_j \nabla h_j(\bar{x}) = 0,$$

$$\lambda \geq 0$$

only having the solution $(\lambda, \mu) = (0, 0)$).

(c) M satisfies the *Slater constraint qualification (SCQ)*, if the vectors $\nabla h_j(x), j \in J$, are linearly independent for all $x \in M$ and if there exists some $x^\star \in \mathbb{R}^n$ with $g_i(x^\star) < 0, i \in I, h_j(x^\star) = 0, j \in J$. The point $x^\star$ is then called a *Slater point*.

The first two constraint qualifications in Definition 2.7.20 fundamentally differ from the third in that the LICQ and the MFCQ are *local* conditions at *one* point $\bar{x} \in M$, while the SCQ is a *global* condition on all of M.

Exercise 2.7.21 Show that the LICQ at $\bar{x} \in M$ implies the MFCQ at $\bar{x} \in M$, but that the reverse of this statement does not hold.

Lemma 2.7.22 *Let* $h_j(x) = a_j^T x + b_j$, $j \in J$, *and let*

$$A = \begin{pmatrix} a_1^T \\ \vdots \\ a_q^T \end{pmatrix}, \quad b = \begin{pmatrix} b_1 \\ \vdots \\ b_q \end{pmatrix}, \quad \text{thus} \quad h(x) = \begin{pmatrix} h_1(x) \\ \vdots \\ h_q(x) \end{pmatrix} = Ax + b.$$

Then M satisfies the SCQ if and only if the rank of A is q and if some $x^* \in \mathbb{R}^n$ exists with $g_i(x^*) < 0$, $i \in I$, and $Ax^* + b = 0$.

Proof For all $j \in J$ the gradients are $\nabla h_j(x) = a_j$, and they are independent of x. □

The following two results are proven in [37]. In Theorem 2.7.23 it may seem surprising at first glance that it postulates the equivalence of the local MFCQ with the global SCQ for convexly described sets. For its proof, however, it is central that convexity itself is a global requirement.

Theorem 2.7.23 *Let the functions g_i, $i \in I$, be convex and C^1 on $\mathbb{R}^n$, let the functions h_j, $j \in J$, be linear, and let $M \neq \emptyset$. Then the following statements are equivalent:*

(a) The MFCQ holds somewhere in M.
(b) The MFCQ holds everywhere in M.
(c) M satisfies the SCQ.

For the following theorem, no convexity requirement is necessary.

Theorem 2.7.24 (Karush-Kuhn-Tucker Theorem for Nonlinear Problems) *Let the optimization problem P be C^1, and let $\bar{x} \in M$ be a local minimal point of P, at which the MFCQ holds. Then $\bar{x}$ is a KKT point of P.*

2.7 Optimality Conditions for Constrained Convex Problems

Because of Exercise 2.7.21, in Theorem 2.7.24 one can also assume the easier to verify, but stronger LICQ, instead of the MFCQ. In convex problems, one may instead switch from the MFCQ to the easier to verify SCQ.

Theorem 2.7.25 (Karush-Kuhn-Tucker Theorem for Convex Problems)
Let the optimization problem P be convexly described and C^1, let M satisfy the SCQ, and let $\bar{x} \in M$ be a minimal point of P. Then $\bar{x}$ is a KKT point of P.

Proof Theorems 2.7.23 and 2.7.24. □

Theorem 2.7.25 implies in particular that, under the SCQ in M, in the case of solvability of P the duality gap vanishes.

In summary, we have shown the following relationships for convex optimization problems P.

For every unconstrained convex C^1-problem P, the following holds:

$$\bar{x} \text{ critical point} \quad \underset{\text{Theorem 2.4.2}}{\overset{\text{Theorem 2.4.5}}{\underset{\Leftarrow}{\Rightarrow}}} \quad \bar{x} \text{ global minimal point.}$$

For every constrained convexly described C^1-problem P, the following holds:

$$\bar{x} \text{ KKT point} \quad \underset{\text{SCQ, Theorem 2.7.25}}{\overset{\text{Theorem 2.7.15}}{\underset{\Leftarrow}{\Rightarrow}}} \quad \bar{x} \text{ global minimal point.}$$

While in the unconstrained case the characterization of global minimal points as critical points is obtained without additional conditions (Corollary 2.4.6), in the constrained case the Slater constraint qualification is needed for a part of the characterization.

Corollary 2.7.26 (Characterization of Global Minimal Points Under SCQ) *Let the optimization problem P be convexly described and C^1, and let M satisfy the SCQ. Then the global minimal points of P are exactly the KKT points of P.*

Proof Theorems 2.7.15 and 2.7.25. □

Since the SCQ is often satisfied in convexly described optimization problems, it does not pose an overly restrictive extra assumption for the characterization statement in Corollary 2.7.26.

We remark that the characterization of global minimal points by KKT points from Corollary 2.7.26 even holds in the 'hidden convex case' where a convex set M is described by inequality constraints with nonconvex C^1-functions g_i, $i \in I$, as long as the SCQ and some mild nondegeneracy assumption are satisfied [27]. The proof of this result partly relies on concepts from convex analysis and would thus go beyond the scope of this textbook.

The situation simplifies if all equality and inequality constraints are linear. In this case, it is not necessary to assume a constraint qualification in the Karush-Kuhn-Tucker theorem [37].

Theorem 2.7.27 (Karush-Kuhn-Tucker Theorem for Linear Constraints) *Let the function f be C^1 on $\mathbb{R}^n$, let the functions g_i, $i \in I$, h_j, $j \in J$, be linear, and let $\bar{x} \in M$ be a local minimal point of P. Then $\bar{x}$ is a KKT point of P.*

Corollary 2.7.28 (Characterization of Global Minimal Points for Linear Constraints) *Let the function f be convex and C^1 on $\mathbb{R}^n$, and let the functions g_i, $i \in I$, h_j, $j \in J$, be linear. Then the global minimal points of P are exactly the KKT points of P.*

Proof Theorems 2.7.15 and 2.7.27. □

Corollary 2.7.29 (Characterization of Global Minimal Points in LPs) *Let P be a linear optimization problem. Then the global minimal points of P are exactly the KKT points of P.*

2.7 Optimality Conditions for Constrained Convex Problems

Proof P fulfills all assumptions of Corollary 2.7.28, because the linear objective function f is convex and C^1 on $\mathbb{R}^n$. □

Corollary 2.7.29 is intimately related to the simplex algorithm of linear optimization, because its termination criterion is fulfilled if and only if it has identified a KKT point.

We conclude this section with an example illustrating the application of the Karush-Kuhn-Tucker conditions.

Example 2.7.30 We consider the problem of determining the distance of an arbitrary point $z \in \mathbb{R}^2$ to the set

$$M = \{x \in \mathbb{R}^2 \mid x_1^2 + x_2^2 \leq 1, \ x_2 \geq 0\},$$

that is, the optimal value of the projection problem

$$P: \quad \min_x \|x - z\|_2 \quad \text{s.t.} \quad x_1^2 + x_2^2 \leq 1,$$

$$x_2 \geq 0.$$

Figure 2.17 illustrates the geometric situation.

Since the solution of P depends on the position of the parameter z, P is a parametric optimization problem [35], which could also be more precisely denoted as $P(z)$. We will ignore this dependence in the following.

By squaring the objective function, we obtain the convexly described C^1-problem

$$Q: \quad \min_x f(x) = x^\mathsf{T} x - 2z^\mathsf{T} x + z^\mathsf{T} z \quad \text{s.t.} \quad g_1(x) = x_1^2 + x_2^2 - 1 \leq 0,$$

$$g_2(x) = -x_2 \leq 0,$$

whose feasible set M satisfies the SCQ. The distance from z to M (i.e., the minimal value of P) coincides with the square root of the minimal value of Q, and the global minimal points of P are identical to those of Q. According to Corollary 2.7.26, the global minimal points of Q are also exactly the KKT points of Q. To calculate the

Fig. 2.17 Distance problem

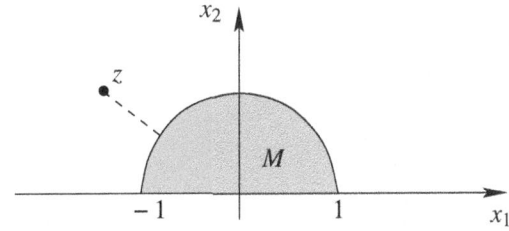

distance from z to M, it is therefore sufficient to evaluate the objective function of P at the KKT points of Q.

First, we provide the gradients of the involved functions:

$$\nabla f(x) = 2(x-z), \quad \nabla g_1(x) = 2x, \quad \nabla g_2(x) = \begin{pmatrix} 0 \\ -1 \end{pmatrix}.$$

Let x be a KKT point of Q. In any case, x then satisfies the two inequalities defining M, and in addition, to formulate the KKT conditions, one must either set up the complementarity conditions or alternatively perform a case distinction according to the possible active index sets. Since the index set $I = \{1, 2\}$ consists of two elements, it has four different subsets. We next consider these case distinctions.

Case 1: $I_0(x) = \emptyset$
The KKT system is

$$0 = \nabla f(x) = 2(x-z),$$
$$1 > x^T x,$$
$$0 < x_2.$$

From this follow the equation $x = z$ as well as the inequalities $z^T z < 1$ and $z_2 > 0$. This also makes it clear for which z this KKT system is solvable at all, namely for

$$z \in M_< := \left\{ x \in \mathbb{R}^n \mid x_1^2 + x_2^2 < 1, \ x_2 > 0 \right\}.$$

The solution of the KKT system and thus the global minimal point of Q in this case is

$$x^\star = z,$$

and the minimal value, i.e., the distance from z to M, is zero.

Case 2: $I_0(x) = \{1\}$
The KKT system is

$$0 = \nabla f(x) + \lambda_1 \nabla g_1(x) = 2(x-z) + \lambda_1 2x,$$
$$1 = x^T x,$$
$$0 < x_2,$$
$$0 \leq \lambda_1.$$

2.7 Optimality Conditions for Constrained Convex Problems

From the first equation follows $(1 + \lambda_1)x = z$. To use the second equation, we form the inner product of the vector with itself on both sides and obtain

$$(1+\lambda_1)^2 \underbrace{x^T x}_{=1} = z^T z \quad \Rightarrow \quad 1 + \lambda_1 = \pm \|z\|_2 \quad \Rightarrow \quad \lambda_1 = \pm \|z\|_2 - 1.$$

Due to $\lambda_1 \geq 0$ only the '+'-alternative is relevant in this condition, and it is solvable exactly for $\|z\|_2 \geq 1$. To determine the associated x, we set $\lambda_1 = \|z\|_2 - 1$ in the equation $(1 + \lambda_1)x = z$ and obtain

$$x^* = \frac{z}{\|z\|_2}.$$

From the condition $x_2^* > 0$ follows $z_2 > 0$. The minimal value of P is

$$\|x^* - z\|_2 = \left\|\left(\frac{1}{\|z\|_2} - 1\right)z\right\|_2 = \left(1 - \frac{1}{\|z\|_2}\right)\|z\|_2 = \|z\|_2 - 1.$$

In summary, in the case $\|z\|_2 \geq 1$, $z_2 > 0$ the point $x^* = z/\|z\|_2$ is the global minimal point with $\text{dist}(z, M) = \|z\|_2 - 1$.

Case 3: $I_0(x) = \{2\}$
The KKT system is

$$0 = \nabla f(x) + \lambda_2 \nabla g_2(x) = 2(x - z) + \lambda_2 \begin{pmatrix} 0 \\ -1 \end{pmatrix},$$

$$1 > x^T x,$$

$$0 = x_2,$$

$$0 \leq \lambda_2.$$

From the equations of this system follows $x_1 = z_1$, $x_2 = 0$ and $\lambda_2 = -2z_2$, in particular

$$x^* = \begin{pmatrix} z_1 \\ 0 \end{pmatrix}.$$

The minimal value is therefore $\|x^* - z\| = \|(0, -z_2)^T\| = |z_2|$.

The inequalities provide information about which z the system is solvable for: From $1 > (x^*)^T x^* = z_1^2$ it follows $z_1 \in (-1, 1)$, and from $0 \leq \lambda_2 = -2z_2$ it follows $z_2 \leq 0$. In summary, for all z with $z_1 \in (-1, 1)$ and $z_2 \leq 0$ the minimal point $x^* = (z_1, 0)^T$ with $\text{dist}(z, M) = -z_2$ is obtained.

Case 4: $I_0(x) = \{1, 2\}$
The KKT system is

$$0 = \nabla f(x) + \lambda_1 \nabla g_1(x) + \lambda_2 \nabla g_2(x) = 2(x - z) + 2\lambda_1 x + \lambda_2 \begin{pmatrix} 0 \\ -1 \end{pmatrix},$$

$$1 = x^T x,$$

$$0 = x_2,$$

$$0 \leq \lambda_1,$$

$$0 \leq \lambda_2.$$

From the second and third equation, the two solution candidates

$$x^1 = \begin{pmatrix} -1 \\ 0 \end{pmatrix} \quad \text{and} \quad x^2 = \begin{pmatrix} 1 \\ 0 \end{pmatrix}$$

follow. For x^1, the first equation yields

$$z = (1 + \lambda_1) \begin{pmatrix} -1 \\ 0 \end{pmatrix} + \frac{\lambda_2}{2} \begin{pmatrix} 0 \\ -1 \end{pmatrix} = -\begin{pmatrix} 1 + \lambda_1 \\ \frac{\lambda_2}{2} \end{pmatrix}$$

and thus $\lambda_1 = -z_1 - 1$ and $\lambda_2 = -2z_2$. From the sign conditions on λ_1 and λ_2, the optimality of x^1 for $z_1 \leq -1$, $z_2 \leq 0$ with $\text{dist}(z, M) = \|x^1 - z\|_2 = \sqrt{(z_1 + 1)^2 + z_2^2}$ is derived. Analogously, one convinces oneself of the optimality of x^2 in the case $z_1 \geq 1$, $z_2 \leq 0$, with $\text{dist}(z, M) = \|x^2 - z\|_2 = \sqrt{(z_1 - 1)^2 + z_2^2}$.

As a summary of the results in all four cases, we obtain

$$\text{dist}(z, M) = \begin{cases} 0, & \text{if } \|z\|_2 < 1, \ z_2 > 0 \\ \|z\|_2 - 1, & \text{if } \|z\|_2 \geq 1, \ z_2 > 0 \\ -z_2, & \text{if } |z_1| < 1, \ z_2 \leq 0 \\ \sqrt{(z_1 + 1)^2 + z_2^2}, & \text{if } z_1 \leq -1, \ z_2 \leq 0 \\ \sqrt{(z_1 - 1)^2 + z_2^2}, & \text{if } z_1 \geq 1, \ z_2 \leq 0. \end{cases}$$

Figure 2.18 shows some level lines of the function $z \mapsto \text{dist}(z, M)$.

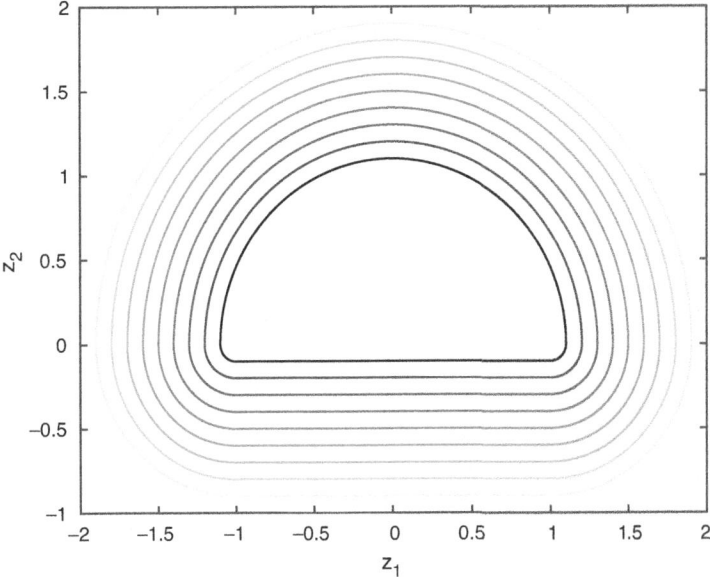

Fig. 2.18 Level lines of dist(z, M) in Example 2.7.30

2.8 Algorithms

According to Theorem 2.1.6, to identify *global* minimal points of convex optimization problems, methods that only generate *local* minimal points are in principle sufficient. Such methods are not developed in global, but in nonlinear (local) optimization [37] and are therefore only briefly outlined here. When applying these methods, due to their different smoothness requirements, the defining functions of P must be sufficiently often continuously differentiable.

In fact, the methods of nonlinear optimization often do not guarantee the identification of a local minimal point, but use a first-order optimality condition as a termination criterion, thus generating the approximation of a critical point for unconstrained problems or of a KKT point for constrained problems. Fortunately, also this is sufficient for convexly described optimization problems, because we have

- according to Theorem 2.4.5 for every unconstrained convex C^1-problem P, that every algorithm that generates a critical point $x^\star$, identifies with $x^\star$ also a global minimal point (e.g., gradient method, (quasi-)Newton method, CG method, trust-region method [37]), and
- according to Theorem 2.7.15 for every constrained convexly described C^1-problem P, that every algorithm that generates a KKT point $x^\star$, identifies with $x^\star$

also a global minimal point (e.g., penalty method, barrier method, SQP method [37]).

For completeness, Sects. 2.8.1 and 2.8.2 briefly present the basic ideas of two central methods of nonlinear unconstrained optimization, namely the gradient method and the Newton method. For details and more advanced methods, refer to [37]. While these methods ignore the structural peculiarity of a convex optimization problem and benefit at most indirectly from convexity, we subsequently present three methods tailored to convexity.

Section 2.8.3 first briefly discusses the idea of cutting planes for an unconstrained convex problem, before in Sect. 2.8.4 Kelley's cutting plane method for constrained convex problems is presented. A different approach to solving constrained convex problems is pursued by the Frank-Wolfe method discussed in Sect. 2.8.5. While the basic ideas of these two older methods can be found in many modern algorithms for larger problem classes (such as in mixed-integer optimization [36]), for convexly described C^1-problems they are typically outperformed by the primal-dual interior point methods described in Sect. 2.8.6, which explicitly exploit duality information. For *nonsmooth* convex problems, there exist several methods (e.g., subgradient and bundle methods), which we cannot cover in the context of this textbook [3].

2.8.1 Basic Idea of the Gradient Method

The gradient method is an iterative method which improves a user-specified starting point $x^0 \in \mathbb{R}^n$ step by step until the gradient of the objective function f at the current iterate is 'sufficiently short'. In this sense, the last generated iterate is the approximation of a critical point of the objective function.

To implement this, in addition to x^0 the user specifies some termination tolerance $\varepsilon > 0$. In the first iteration, it is checked whether $\|\nabla f(x^0)\| \leq \varepsilon$ coincidentally already holds. In this case, the procedure stops with x^0 as the sought-after approximation of a critical point.

Otherwise, one uses the fact that $\nabla f(x^0)$ stands perpendicular to the level set $\{x \in \mathbb{R}^n \mid f(x) = f(x^0)\}$ and points in the direction of steepest *ascent* of f (Fig. 2.19). To minimize f starting from x^0, one can therefore take a step in the direction of steepest *descent*, $d^0 = -\nabla f(x^0)$. The step size $t^0 > 0$ is determined with the help of a step size control, such as Armijo's rule [37]. One then sets $x^1 = x^0 + t^0 d^0$ and checks again $\|\nabla f(x^1)\| \leq \varepsilon$ etc.

Fig. 2.19 Gradient and level set of f at x^0

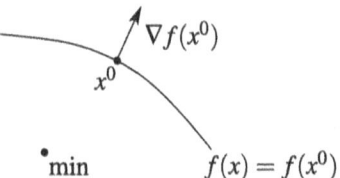

2.8 Algorithms

Under mild assumptions the gradient method terminates after a finite number of steps. However, this can take a long time due to the so-called zigzagging effect, which persists also for convex functions [37].

2.8.2 Basic Idea of the Newton Method

The Newton method was initially designed for finding zeros of functions. To approximate a solution of the equation $g(x) = 0$ for a C^1-function $g : \mathbb{R}^n \to \mathbb{R}^n$, it proceeds as follows: Again the user provides a starting point $x^0 \in \mathbb{R}^n$ and a termination tolerance $\varepsilon > 0$. If $\|g(x^0)\| \leq \varepsilon$ coincidentally already holds, the procedure stops with x^0 as an approximation of a zero.

Otherwise, g is *linearized* around x^0 and the linearized problem is solved. For the case $n = 1$, the procedure is illustrated in Fig. 2.20. 'Linearizing' g in this illustration means that the graph of g is replaced by its tangent at the point $(x^0, g(x^0))$, i.e., g is approximated by the function $g(x^0) + g'(x^0)(x - x^0)$. 'Solving the linearization' means that a zero of the linear function describing the tangent is determined: From

$$0 = g(x^0) + g'(x^0)(x - x^0)$$

one obtains as the new iterate the point

$$x^1 = x^0 - \frac{g(x^0)}{g'(x^0)}.$$

For this, of course, $g'(x^0) \neq 0$ must hold.

For general n, finding a zero for the linearization means solving the linear system of equations

$$0 = g(x^0) + \underbrace{Dg(x^0)}_{(n,n)\text{-matrix}} (x - x^0),$$

Fig. 2.20 Newton method for finding a zero

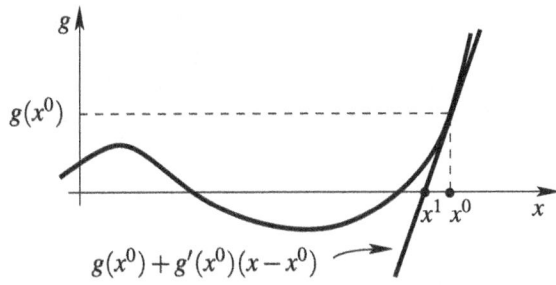

from which

$$x^1 = x^0 - (Dg(x^0))^{-1} g(x^0)$$

follows. For this, the Jacobian matrix $Dg(x^0)$ must be invertible. Then one again checks $\|g(x^1)\| \leq \varepsilon$, and so on.

Optimization problems are solved via the Newton method by looking for a critical point, i.e., the problem $g(x) := \nabla f(x) = 0$ is solved by the Newton method. In iteration k one obtains

$$x^{k+1} = x^k - (D^2 f(x^k))^{-1} \nabla f(x^k).$$

Along the search direction $d^k = -(D^2 f(x^k))^{-1} \nabla f(x^k)$ one can additionally determine a step size $t^k > 0$ (e.g., again by Armijo's rule) and then define $x^{k+1} = x^k + t^k d^k$ (in this case, one speaks of the *damped* Newton method). The convergence of this method is very fast if x^0 lies close to a solution.

Disadvantages of the Newton method in the general case are:

- One usually does not know if x^0 is close enough to a critical point.
- $D^2 f(x^k)$ is not necessarily invertible (i.e., $D^2 f(x^k) d^k = -\nabla f(x^k)$ is not necessarily solvable).
- Any critical points are approximated, including local maximal points and saddle points. This is not to be expected in the gradient method, as the choices of search direction and step size reduce the objective function value in each iteration.

The *last* disadvantage of the Newton method does not occur in convex problems, as for them according to Theorem 2.4.5 all critical points are global minimal points anyway. The convergence of the Newton method to a critical point is therefore sufficient for convergence to a global minimal point.

2.8.3 Basic Idea of Cutting Plane Methods

As a first class of methods tailored to convexity, we consider cutting plane methods. This section describes the basic idea initially for unconstrained problems. Historically, the reason for introducing cutting plane methods was that *linear* optimization problems were algorithmically solvable since the advent of the simplex algorithm, so that one could try to approximate convex problems by linear ones.

The crucial observation is that, according to Theorem 2.2.3, the graph of a convex C^1-function f lies above each of its tangents (resp. above each of its tangent spaces, if $n > 1$).

2.8 Algorithms

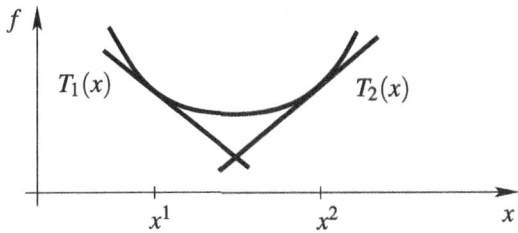

Fig. 2.21 Idea of the cutting plane method for $n = 1$

For two points $x^1, x^2 \in \mathbb{R}^n$, as illustrated in Fig. 2.21 for $n = 1$, the graph of f lies above the graph of the linearization $T_1(x) = f(x^1) + \langle \nabla f(x^1), x - x^1 \rangle$ as well as above that of $T_2(x) = f(x^2) + \langle \nabla f(x^2), x - x^2 \rangle$. For all $x \in \mathbb{R}^n$ it therefore applies

$$f(x) \geq \max\{T_1(x), T_2(x)\}.$$

Analogously one may proceed for k points $x^1, \ldots, x^k$ so that f is approximated from below by a piecewise linear function. Instead of minimizing f, one solves the approximating auxiliary problem

$$\min_{x} \max_{i=1,\ldots,k} T_i(x).$$

If the computed minimal point x^{k+1} does not meet a termination criterion (see below), one adds the linearization

$$T_{k+1}(x) = f(x^{k+1}) + \langle \nabla f(x^{k+1}), x - x^{k+1} \rangle$$

to the piecewise linear approximation, solves again, etc.

The solution of the actually nonsmooth auxiliary problem is achieved by epigraph reformulation (Exercise 1.3.7), according to which the following optimization problems are equivalent to each other:

$$\min_{x} \max_{i=1,\ldots,k} T_i(x) \Leftrightarrow \min_{x,\alpha} \alpha \text{ s.t. } \max_{i=1,\ldots,k} T_i(x) \leq \alpha$$

$$\Leftrightarrow \min_{x,\alpha} \alpha \text{ s.t. } T_i(x) \leq \alpha, \ i = 1, \ldots, k$$

$$\Leftrightarrow \min_{x,\alpha} \alpha \text{ s.t. } \underbrace{f(x^i) + \langle \nabla f(x^i), x - x^i \rangle - \alpha}_{\text{linear in } (x, \alpha)} \leq 0,$$

$$i = 1, \ldots, k.$$

Thus, the approximating auxiliary problem is reformulated into an equivalent linear optimization problem, which can be solved, for example, by the simplex algorithm.

With regard to an appropriate termination criterion, note that an optimal point (x^{k+1}, α^{k+1}) of the above linear problem provides an *enclosure* for the sought optimal value $v = \min_{x \in \mathbb{R}^n} f(x)$. Indeed, in addition to the upper bound $v \le f(x^{k+1})$, the minimality of x^{k+1} also yields for all $x \in \mathbb{R}^n$

$$\alpha^{k+1} = \max_{i=1,\dots,k} T_i(x^{k+1}) \le \max_{i=1,\dots,k} T_i(x) \le f(x),$$

so $\alpha^{k+1} \le v$. The optimal value v therefore lies in the interval $[\alpha^{k+1}, f(x^{k+1})]$, and the termination criterion $f(x^{k+1}) - \alpha^{k+1} \le \varepsilon$ guarantees ε-accuracy in the computation of v.

2.8.4 Kelley's Cutting Plane Method

The cutting plane method published by Kelley in 1960 [25] uses similar ideas, but for *constrained* problems of the form

$$P: \quad \min c^\mathsf{T} x \quad \text{s.t.} \quad g_i(x) \le 0, \ i \in I \ (= \{1, \dots, p\}),$$
$$Ax \le b$$

with convex C^1-functions g_i, $i \in I$, defined on $\mathbb{R}^n$. Every convexly described optimization problem can be brought into this form, because

- if the objective function is nonlinear convex, one 'shifts' it into the constraints by epigraph reformulation,
- if there are equality constraints, these are linear and can therefore each be written as two linear inequality constraints (this technique is only used for linear equality constraints, because for *non*linear ones it destroys important regularity properties [37]).

The distinction between linear and nonlinear convex constraints is important in the following, as linear constraints do not need to be linearized. We therefore define the sets

$$K = \{x \in \mathbb{R}^n | \ g_i(x) \le 0, \ i \in I\}$$

and

$$L = \{x \in \mathbb{R}^n | \ Ax \le b\}.$$

The feasible set is thus $M := K \cap L$. Furthermore, let M be nonempty and bounded (so that P has a global minimal point according to the Weierstrass theorem).

2.8 Algorithms

As an accompanying example, we solve the problem

$$P: \quad \min x_2 \quad \text{s.t.} \quad x_1^2 + x_2^2 \leq 1,$$
$$x_2 \geq e^{x_1},$$
$$x_1 \leq -\tfrac{1}{2},$$
$$x_2 \geq x_1.$$

Here we have

$$c = \begin{pmatrix} 0 \\ 1 \end{pmatrix},$$

$$g_1(x) = x_1^2 + x_2^2 - 1,$$
$$g_2(x) = e^{x_1} - x_2,$$

$$A = \begin{pmatrix} 1 & 0 \\ 1 & -1 \end{pmatrix},$$

$$b = \begin{pmatrix} -\tfrac{1}{2} \\ 0 \end{pmatrix},$$

$$K = \left\{ x \in \mathbb{R}^2 \mid x_1^2 + x_2^2 \leq 1, \ x_2 \geq e^{x_1} \right\},$$
$$L = \left\{ x \in \mathbb{R}^2 \mid x_1 \leq -\tfrac{1}{2}, \ x_2 \geq x_1 \right\}.$$

The feasible set of P is shown in Fig. 2.22.

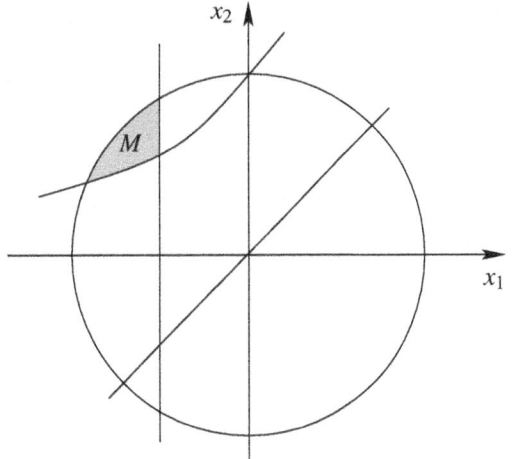

Fig. 2.22 Feasible set of P in the accompanying example

The basic idea of the method is to approximate M from the outside by linear inequality constraints and to successively improve this approximation. Since L is already described by linear inequality constraints, the approximation only affects K.

To do this, first determine a convex polyhedron (i.e., a set described by a finite number of linear inequalities) B with $K \subseteq B$ and set $M^0 := B \cap L$. Then M^0 is a convex polyhedron with $M \subseteq M^0$. Due to the boundedness of M, B can be chosen so that M^0 becomes a convex polytope, i.e., a nonempty and bounded convex polyhedron (we remark that some authors allow convex polytopes to be empty).

B may be a very rough approximation of K, the simplest is often a cuboid (also called *box*; Sect. 3.3). Even without a geometric idea about the shape of the set K, such a box can often be formally constructed. In the accompanying example, one can proceed as follows:

$$x \in K \quad \Rightarrow \quad 1 \geq x_1^2 + x_2^2 \geq x_1^2 \quad \Rightarrow \quad x_1 \in [-1, 1]$$

and analogously $x_2 \in [-1, 1]$. It follows

$$K \subseteq B := [-1, 1]^2$$

and thus

$$M^0 = B \cap L = \left\{ x \in [-1, 1]^2 \mid x_1 \leq -\tfrac{1}{2},\ x_2 \geq x_1 \right\}.$$

Figure 2.23 shows the set M^0.

The auxiliary problem

$$LP^0: \quad \min c^T x \quad \text{s.t.} \quad x \in M^0$$

Fig. 2.23 Initial set M^0

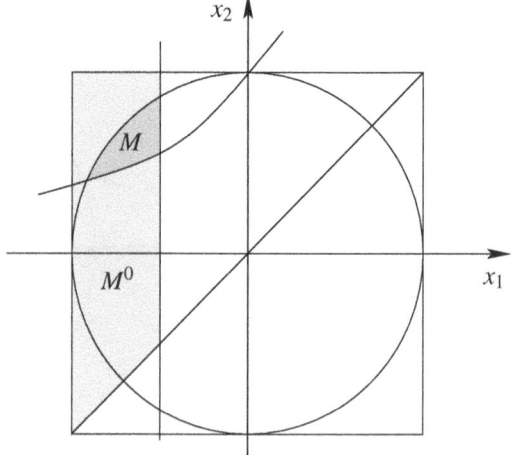

2.8 Algorithms

can be solved, for example, using the simplex algorithm. Let x^0 be a minimal point of LP^0. From

$$c^\mathsf{T} x \geq c^\mathsf{T} x^0 \quad \forall x \in M^0$$

and $M \subseteq M^0$ follows

$$c^\mathsf{T} x \geq c^\mathsf{T} x^0 \quad \forall x \in M.$$

If x^0 lies in M, then x^0 is also a minimal point of P.

In the example, we obtain

$$x^0 = \begin{pmatrix} -1 \\ -1 \end{pmatrix} \notin M,$$

so x^0 is not an optimal point of P.

In the case $x^0 \notin M$ because of $M^0 \subseteq L$ (at least) one of the inequalities in K must be violated, i.e., it holds $\max_{i \in I} g_i(x^0) > 0$. Choose one of the most violated inequalities, i.e., some $\ell \in I$ with $g_\ell(x^0) = \max_{i \in I} g_i(x^0)$. In the example, it holds $g_1(x^0) = 1$ and $g_2(x^0) = \frac{1}{e} + 1$, so $\ell = 2$.

The resulting *cut*, which lends the cutting plane method its name, is

$$M^1 = M^0 \cap \left\{ x \in \mathbb{R}^n \mid g_\ell(x^0) + \langle \nabla g_\ell(x^0), x - x^0 \rangle \leq 0 \right\}.$$

The new set M^1 has three important properties:

- It holds $x^0 \notin M^1$, because $g_\ell(x^0) + \langle \nabla g_\ell(x^0), x^0 - x^0 \rangle = g_\ell(x^0) > 0$, i.e., the old optimal point x^0 is 'cut off' and cannot reappear in future iterations.
- It holds $M \subseteq M^1$, because

$$\forall x \in M: \quad g_\ell(x^0) + \langle \nabla g_\ell(x^0), x - x^0 \rangle \overset{\text{Theorem 2.2.3}}{\leq} g_\ell(x) \leq 0.$$

- M^1 is again a convex polytope.

In the example, the new inequality for M^1 is obtained as follows:

$$0 \geq g_2(x^0) + \langle \nabla g_2(x^0), x - x^0 \rangle = \tfrac{1}{e} + 1 + (\tfrac{1}{e}, -1) \begin{pmatrix} x_1 + 1 \\ x_2 + 1 \end{pmatrix}$$

$$= \tfrac{1}{e} + 1 + \tfrac{x_1}{e} + \tfrac{1}{e} - x_2 - 1 = \tfrac{x_1}{e} - x_2 + \tfrac{2}{e}.$$

In Fig. 2.24, the set M^1 is shown.

Fig. 2.24 The set M^1

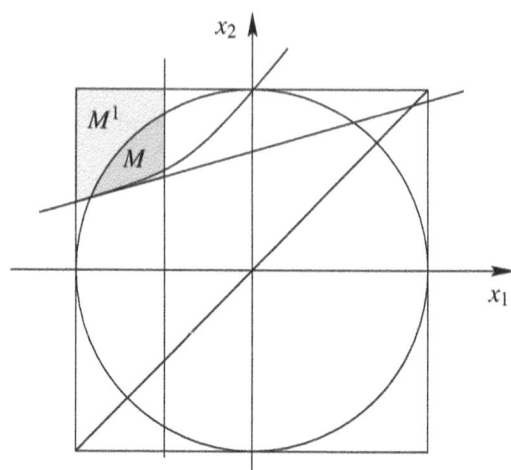

Next, solve

$$LP^1: \quad \min c^\mathsf{T} x \quad \text{s.t.} \quad x \in M^1$$

etc., until the optimal point x^k of a subproblem LP^k lies in M. Due to the outer approximation of M by $M^k \supsetneq M$ for all k, one cannot expect that any iterate x^k actually lies in M (since the optimal points x^k typically lie on the boundary of M^k). Therefore the termination criterion is relaxed to $\max_{i \in I} g_i(x^k) \leq \varepsilon$ with a tolerance $\varepsilon > 0$, and one already accepts such *ε-feasible minimal points* (see also Sect. 3.9). The complete procedure is given in Algorithm 2.1.

It is clear that Algorithm 2.1 generates a sequence (x^k) that approaches M from the outside. However, for the method to terminate after a finite number of steps for any tolerance $\varepsilon > 0$, one must exclude that the sequence (x^k) maintains a positive distance to M, in the sense that the values $\max_{i \in I} g_i(x^k)$ for $k \to \infty$ do not drop below any positive number.

Theorem 2.8.1 *Algorithm 2.1 terminates after a finite number of steps.*

Proof Assume that the algorithm does not terminate. Then the termination criterion in line 5 is not met for any $k \in \mathbb{N}$, i.e., the procedure generates an infinite sequence (x^k). In particular, infinitely often some $\ell_k \in I$ is chosen in line 6. Due to $|I| < \infty$ at least one index $\ell \in I$ occurs infinitely often. We consider the subsequence formed by the x^k that correspond to these cutting planes and call them again (x^k), to avoid a subsequence notation.

2.8 Algorithms

Algorithm 2.1: Kelley's cutting plane method

Input: Continuously differentiable convexly described minimization problem P with objective function $c^\mathsf{T} x$ and nonempty, bounded feasible set $M = K \cap L$, termination tolerance $\varepsilon > 0$

Output: ε-feasible minimal point $\bar{x}$ of P, i.e. $\bar{x}$ with $\max_{i \in I} g_i(\bar{x}) \leq \varepsilon$ and $c^\mathsf{T} \bar{x} \leq c^\mathsf{T} x$ for all $x \in M$

1 **begin**
2   Choose a convex polyhedron B with $K \subseteq B$, so that $M^0 := B \cap L$ is a convex polytope.
3   Determine a minimal point x^0 of

$$LP^0: \quad \min c^\mathsf{T} x \quad \text{s.t.} \quad x \in M^0.$$

4   Set $k = 0$.
5   **while** $\max_{i \in I} g_i(x^k) > \varepsilon$ **do**
6   Choose some $\ell \in I$ with $g_\ell(x^k) = \max_{i \in I} g_i(x^k)$.
7   Set

$$M^{k+1} = M^k \cap \left\{ x \in \mathbb{R}^n \mid g_\ell(x^k) + \langle \nabla g_\ell(x^k), x - x^k \rangle \leq 0 \right\}.$$

8   Replace k with $k + 1$.
9   Determine a minimal point x^k of

$$LP^k: \quad \min c^\mathsf{T} x \quad \text{s.t.} \quad x \in M^k.$$

10   **end**
11   Set $\bar{x} = x^k$.
12 **end**

Since the sequence (x^k) lies in the compact set $B \cap L$, it possesses a convergent subsequence. We switch to such a further subsequence, still call it (x^k) and denote its limit point with $x^\star$. Then $g_\ell(x^k) > \varepsilon$ holds for all $k \in \mathbb{N}$ and therefore, after taking the limit, $g_\ell(x^\star) \geq \varepsilon$.

Because of $x^{k+1} \in M^{k+1} \subseteq M^k$ (also after forming the above subsequences) it follows

$$g_\ell(x^k) + \langle \nabla g_\ell(x^k), x^{k+1} - x^k \rangle \leq 0,$$

and taking the limit $k \to \infty$ yields

$$g_\ell(x^\star) + \underbrace{\langle \nabla g_\ell(x^\star), x^\star - x^\star \rangle}_{=0} \leq 0,$$

contradicting $g_\ell(x^\star) \geq \varepsilon$. Thus, the assumption was wrong, and the algorithm terminates. □

2.8.5 The Frank-Wolfe Method

The following method is based on a fundamentally different idea than cutting plane methods: It generates a sequence of *feasible* iterates (x^k), until some x^k satisfies an *optimality condition* ε-approximately, while Kelley's cutting plane method generates a sequence of *optimal* iterates (x^k) until some x^k satisfies a *feasibility measure* ε-approximately.

The method will use a construction from the proof of the subsequent Theorem 2.8.2. Both in this proof and in the algorithmic implementation of the result, we will use the fact (proven e.g. in [37] by Taylor's theorem), that for a point $\bar{x} \in \mathbb{R}^n$ and a C^1-function f, every direction $d \in \mathbb{R}^n$ with $\langle \nabla f(x^0), d \rangle < 0$ is a (first-order) *descent direction* for f at $\bar{x}$, i.e., for all sufficiently small $t > 0$ it holds $f(\bar{x} + td) < f(\bar{x})$.

Theorem 2.8.2 (Variational Formulation of Convex Problems) *Let the problem*

$$P: \quad \min f(x) \quad \text{s.t.} \quad x \in M$$

with nonempty and convex feasible set M and convex objective function $f \in C^1(M, \mathbb{R})$ be given. Then the following statements hold:

(a) A point $\bar{x} \in \mathbb{R}^n$ is a global minimal point of P if and only if $\bar{x}$ is a global minimal point of

$$Q(\bar{x}): \quad \min_x \langle \nabla f(\bar{x}), x - \bar{x} \rangle \quad \text{s.t.} \quad x \in M.$$

(b) Let $\bar{x} \in M$. Then the minimal value of $Q(\bar{x})$ satisfies $v(\bar{x}) \leq 0$, and $\bar{x}$ is a global minimal point of P if and only if $v(\bar{x}) = 0$ holds.

Proof To prove statement a, let $\bar{x}$ be a global minimal point of $Q(\bar{x})$. Then in particular $\bar{x} \in M$ holds, so that $\bar{x}$ is also feasible for P. Furthermore, for all $x \in M$ we have

$$f(x) \overset{\text{Theorem 2.2.3}}{\geq} f(\bar{x}) + \langle \nabla f(\bar{x}), x - \bar{x} \rangle \geq f(\bar{x}) + \langle \nabla f(\bar{x}), \bar{x} - \bar{x} \rangle = f(\bar{x}),$$

so $\bar{x}$ is a global minimal point of P.

On the other hand, let $\bar{x}$ be a global minimal point of P. In particular, $\bar{x}$ is then feasible for $Q(\bar{x})$. Let us assume that $\bar{x}$ is not a global minimal point of $Q(\bar{x})$. Then there exists some $y \in M$ with

$$\langle \nabla f(\bar{x}), y - \bar{x} \rangle < \langle \nabla f(\bar{x}), \bar{x} - \bar{x} \rangle = 0.$$

2.8 Algorithms

The direction $d := y - \bar{x}$ is therefore a (first-order) descent direction for f in $\bar{x}$. Moreover, for all sufficiently small steps $t > 0$, one does not leave the feasible set M when moving from $\bar{x}$ along d, because due to $\bar{x}, y \in M$ and the convexity of M, all step sizes $t \in [0, 1]$ satisfy

$$\bar{x} + td = (1 - t)\bar{x} + ty \in M.$$

For each sufficiently small $t > 0$, the point $\bar{x} + td$ is therefore feasible for P and has a strictly smaller objective function value than $\bar{x}$. This contradicts the assumption that $\bar{x}$ is a minimal point of P.

The first claim of statement b follows from $\bar{x} \in M$. Furthermore, from statement a, it follows that every global minimal point $\bar{x}$ of P is also a global minimal point of $Q(\bar{x})$, from which $v(\bar{x}) = \langle \nabla f(\bar{x}), \bar{x} - \bar{x} \rangle = 0$ results. On the other hand, if $\bar{x} \in M$ is not a global minimal point of P, then $\bar{x}$ is not a global minimal point of $Q(\bar{x})$ according to statement a. Since $M \neq \emptyset$, there is some $y \in M$ with $\langle \nabla f(\bar{x}), y - \bar{x} \rangle < \langle \nabla f(\bar{x}), \bar{x} - \bar{x} \rangle = 0$. Due to the feasibility of $\bar{x}$ for $Q(\bar{x})$, it must then hold $v(\bar{x}) < 0$. □

The Frank-Wolfe method is based on solving auxiliary problems of the form $Q(\bar{x})$ from Theorem 2.8.2. For these to be solvable in the first place, we require in the following that M is not only nonempty and convex, but also compact. Furthermore, the method will only be algorithmically interesting if the problems $Q(\bar{x})$ can be solved quickly (as opposed to, for example, applying Kelley's cutting plane method to each auxiliary problem). This will lead to further conditions on M.

The algorithmic approach proceeds as follows: We first choose a starting point $x^0 \in M$ and determine the minimal value v^0 of $Q^0 := Q(x^0)$. If $v^0 = 0$ applies, the procedure terminates because, according to Theorem 2.8.2b, x^0 is a global minimal point of P.

Since it is not to be expected that already x^0 is a global minimal point, the crucial question is how to proceed in the case $v^0 < 0$. For this, a minimal point y^0 of Q^0 is determined along with v^0. As in the proof of Theorem 2.8.2a, the search direction $d^0 := y^0 - x^0$ is then a feasible descent direction for f in x^0. A feasible point with a smaller objective function value can therefore be found in the form $x^0 + td^0$ with a suitable (i.e., not too large) $t \in [0, 1]$.

For *step size control* (i.e., for the choice of the step size t), proceed as follows: Choose t^0 as a minimal point of $f(x^0 + td^0)$ on $[0, 1]$, i.e., solve the one-dimensional problem

$$S(x^0, d^0): \quad \min_t f(x^0 + td^0) \quad \text{s.t.} \quad 0 \le t \le 1.$$

Also $S(x^0, d^0)$ must be quickly solvable for algorithmic implementation, or one is satisfied with an approximation to t^0. After all, $S(x^0, d^0)$ has a convex objective function and a simple feasible set.

Fig. 2.25 Example for the Frank-Wolfe method

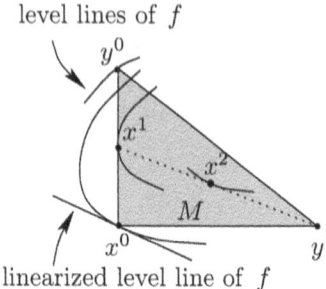

Next set the new iterate to $x^1 := x^0 + t^0 d^0$ and solve Q^1, etc. Figure 2.25 shows the first two iterations of the method in an example. The termination criterion must again be relaxed, as it cannot be expected that $v^k = 0$ will apply after a finite number of steps. Instead, we stop already for $v^k \geq -\varepsilon$ with a given tolerance $\varepsilon > 0$. This implies with Theorem 2.2.3 for the associated iterate x^k and all $x \in M$

$$f(x) \geq f(x^k) + \langle \nabla f(x^k), x - x^k \rangle \geq f(x^k) + v^k \geq f(x^k) - \varepsilon$$

and thus $v \leq f(x^k) \leq v + \varepsilon$. Such a point $x^k \in M$ is called ε-*minimal feasible point*. The complete procedure is given in Algorithm 2.2.

To use it in a sensible manner, the auxiliary problems Q^k and $S(x^k, d^k)$ must be quickly solvable. With respect to Q^k, for example the following conditions are suitable:

- If M is a convex polytope, then Q^k is a linear optimization problem and can be solved, for example, using the simplex algorithm.
- If M is a ball, then y^k can even be calculated in closed form.

For $S(x^k, d^k)$ applies:

- If f is convex-quadratic, then t^k can be calculated in closed form (Exercise 2.8.3).

In view of these considerations, Algorithm 2.2 was formulated in 1956 for convex-quadratic objective functions with linear constraints [11].

Exercise 2.8.3 With $A = A^\mathsf{T} \succ 0$ and $b \in \mathbb{R}^n$ let $f(x) = \frac{1}{2} x^\mathsf{T} A x + b^\mathsf{T} x$. Provide a closed form expression for the unique minimal point t^k of $S(x^k, d^k)$.

The proof of the following theorem is based on techniques of nonlinear optimization [37] and is therefore only outlined here in broad terms. The concept of Lipschitz continuity used in the assumption is treated in detail in Sect. 3.9.1.

2.8 Algorithms

Algorithm 2.2: Frank-Wolfe method

Input: Convex minimization problem P with continuously differentiable objective function and nonempty, compact feasible set M, starting point $x^0 \in M$, termination tolerance $\varepsilon > 0$

Output: ε-minimal feasible point $\bar{x}$ of P, i.e. $\bar{x} \in M$ with $v \leq f(\bar{x}) \leq v + \varepsilon$ (if the algorithm terminates; Theorem 2.8.4)

1 **begin**
2   Set $k = 0$.
3   Determine a minimal point y^0 and the minimal value v^0 of

$$Q^0: \quad \min_x \langle \nabla f(x^0), x - x^0 \rangle \quad \text{s.t.} \quad x \in M.$$

4   **while** $v^k < -\varepsilon$ **do**
5   Set $d^k = y^k - x^k$.
6   Choose t^k as a minimal point of

$$S(x^k, d^k): \quad \min_t f(x^k + td^k) \quad \text{s.t.} \quad 0 \leq t \leq 1.$$

7   Set $x^{k+1} = x^k + t^k d^k$.
8   Replace k with $k + 1$.
9   Determine a minimal point y^k and the minimal value v^k of

$$Q^k: \quad \min_x \langle \nabla f(x^k), x - x^k \rangle \quad \text{s.t.} \quad x \in M.$$

10   **end**
11   Set $\bar{x} = x^k$.
12 **end**

Theorem 2.8.4 *Let the gradient ∇f be Lipschitz continuous on the nonempty, convex and compact set M. Then Algorithm 2.2 terminates after a finite number of steps.*

Sketch of Proof Assume that the algorithm does not terminate. Then $v^k < -\varepsilon$ holds for all $k \in \mathbb{N}$, and infinite sequences (x^k), (y^k) and (t^k) are generated. Since the sequences (x^k) and (y^k) are contained in the compact set M, we may assume them to be convergent without loss of generality (i.e., after possibly taking subsequences).

Due to the step size selection in line 6, (t^k) is an *efficient* sequence of step sizes [37], i.e.,

$$\exists c > 0 \quad \forall k \in \mathbb{N}: \quad f(x^k + t^k d^k) - f(x^k) \leq -c \left(\frac{\langle \nabla f(x^k), d^k \rangle}{\|d^k\|_2} \right)^2$$

holds (this is tedious to show, and here the Lipschitz continuity of ∇f is required). It follows

$$\underbrace{f(x^{k+1}) - f(x^k)}_{<0,\ \to 0} \leq -c\left(\frac{\langle \nabla f(x^k), d^k\rangle}{\|d^k\|_2}\right)^2 \leq 0$$

and therefore by the sandwich theorem

$$\frac{\langle \nabla f(x^k), d^k\rangle}{\|d^k\|_2} \xrightarrow{k\to\infty} 0.$$

The numerator of this fraction cannot converge to zero due to

$$\langle \nabla f(x^k), d^k\rangle = \langle \nabla f(x^k), y^k - x^k\rangle = v^k < -\varepsilon,$$

so its denominator must go to infinity:

$$\|d^k\| \to \infty.$$

However, this is a contradiction, since x^k and y^k stem from the bounded set M and thus the sequence of $d^k = y^k - x^k$ is also bounded. □

Since in the proof of Theorem 2.8.4 only the efficiency of the step size sequence (t^k) is used and not the fact that the t^k are exact global minimal points of $S(x^k, d^k)$, the convergence statement from Theorem 2.8.4 still holds if $S(x^k, d^k)$ is only solved inexactly (e.g., by Armijo's rule) [37].

2.8.6 Basic Idea of Primal-Dual Interior Point Methods

Primal-dual interior point methods are based on a purely primal procedure, namely the *barrier method* (see also [37]). Its basic idea is the approximation of P by unconstrained problems, where a 'barrier' is erected at the boundary bd M of M that enforces feasibility. Let the convexly described C^1-problem

$$P: \quad \min f(x) \quad \text{s.t.} \quad g_i(x) \leq 0, \ i \in I,$$

be given with a bounded feasible set M and

$$M_< := \{x \in \mathbb{R}^n \mid g_i(x) < 0, \ i \in I\} \neq \emptyset$$

(i.e., M has Slater points). The presence of equality constraints (i.e. $J \neq \emptyset$) can also be considered, but for the sake of clarity we omit them. Penalization of points near bd M is implemented by a *barrier function*, i.e., a function β defined on $M_<$ that satisfies $\lim_k \beta(x^k) = +\infty$ for all $(x^k) \subseteq M_<$ with $\lim_k x^k = x^\star \in \text{bd}(M_<)$. Due

2.8 Algorithms

to the boundedness of M, this means that we assume β to be coercive on $M_<$ in the sense of Definition 1.2.40. The constrained problem P is then approximated by the unconstrained barrier problems

$$\min_{x} B(t, x)$$

with the objective function

$$B(t, x) := f(x) + t\beta(x)$$

and *barrier parameters* $t > 0$. The strategy of the barrier method is to successively less penalize the proximity to the boundary of M, i.e., to let the barrier parameter t tend to zero, while tracing minimal points of $B(t, x)$.

For example, for the problem

$$\min -x \quad \text{s.t.} \quad x \leq 0$$

the barrier function $\beta(x) = -\log(-x)$ is sketched in Fig. 2.26, and the behavior of the functions $B(t, x) = -x - t \log(-x)$ for $t \searrow 0$ is illustrated in Fig. 2.27.

For a general feasible set M of P,

$$\beta(x) = -\sum_{i \in I} \log(-g_i(x))$$

Fig. 2.26 Logarithmic barrier function β

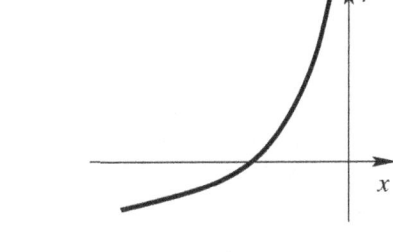

Fig. 2.27 Functions $B(t, x)$ with $t \searrow 0$

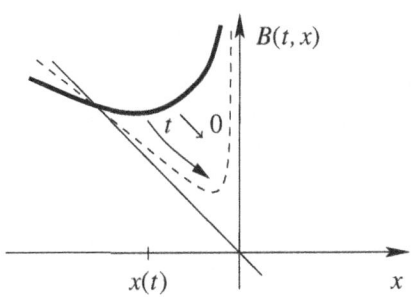

is a barrier function, and barrier problems with objective functions

$$B(t, x) := f(x) - t \sum_{i \in I} \log(-g_i(x))$$

are to be solved for $t \searrow 0$. Under our assumptions, the following result is true.

Theorem 2.8.5 *For all $t > 0$, the function $B(t, \cdot)$ is convex and has a minimal point on its domain $M_<$. If at least one of the functions f, g_i, $i \in I$, is strictly convex on $\mathbb{R}^n$, or if all functions g_i, $i \in I$, are linear, then the minimal point is unique.*

Proof We consider a fixed $t > 0$. With $\gamma(x) := -\log(-x)$, the convexity of the functions $\gamma(g_i)$, $i \in I$, and thus the convexity of $B(t, \cdot) = f + t \sum_{i \in I} \gamma(g_i)$ follow from the convexity and monotonicity of γ on the set of negative numbers. The boundedness of M implies coercivity of $B(t, \cdot)$ on $M_<$, so that the existence of a minimal point is guaranteed by Corollary 1.2.43. Its uniqueness follows with Theorem 2.3.3b, if we can even demonstrate *strict* convexity of the function $B(t, \cdot)$.

Since the functions f and $\gamma(g_i)$, $i \in I$, are convex, the strict convexity of one of these functions is sufficient. Due to the strict monotonicity of γ, the strict convexity of a function g_i also implies the strict convexity of $\gamma(g_i)$, so that the first sufficient condition for strict convexity of $B(t, \cdot)$ is shown.

To see the validity of the second sufficient condition, we show the strict convexity of the barrier function $\beta(x) = \sum_{i \in I} \gamma(a_i^T x - b_i)$ for linear functions $g_i(x) = a_i^T x - b_i$, $i \in I$. Suppose β is not strictly convex on $M_<$. Then there exist $x, y \in M_<$ with $x \neq y$ and $\lambda \in (0, 1)$ with

$$\begin{aligned} 0 &= (1-\lambda)\beta(x) + \lambda\beta(y) - \beta((1-\lambda)x + \lambda y) \\ &= \sum_{i \in I} \left((1-\lambda)\gamma(g_i(x)) + \lambda\gamma(g_i(y)) - \gamma(g_i((1-\lambda)x + \lambda y))\right) \\ &= \sum_{i \in I} \left((1-\lambda)\gamma(a_i^T x - b_i) + \lambda\gamma(a_i^T y - b_i)) - \gamma((1-\lambda)(a_i^T x - b_i)\right. \\ &\quad \left. + \lambda(a_i^T y - b_i))\right). \end{aligned}$$

Since, due to the convexity of the functions $\gamma(g_i)$, $i \in I$, each of the summands is nonnegative, they must all vanish simultaneously. In view of the strict convexity of γ this is only possible for

$$0 = (a_i^T x - b_i) - (a_i^T y - b_i) = a_i^T(x - y), \ i \in I.$$

2.8 Algorithms

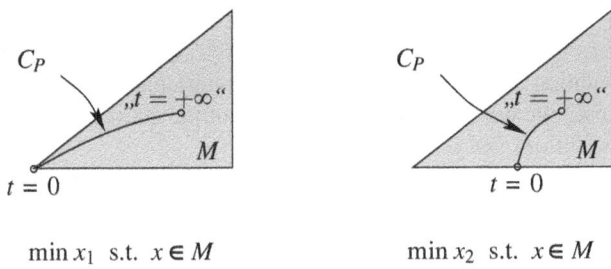

Fig. 2.28 Primal central paths

The vector $x - y \neq 0$ is therefore a recession direction of the set M [36], i.e., from any $z \in M$ steps of arbitrary length $s > 0$ in the direction of $x - y$ do not lead out of M: For all $i \in I$ and $s > 0$ it holds

$$a_i^T(z + s(x - y)) - b_i = a_i^T z - b_i + s a_i^T(x - y) = a_i^T z - b_i \leq 0.$$

This contradicts the assumed boundedness of M. □

Under one of the conditions of Theorem 2.8.5, let $x(t)$ for $t > 0$ denote the unique minimal point of $B(t, \cdot)$. The set

$$C_P = \{x(t) |\, t \in (0, +\infty)\}$$

is called the *primal central path* of P. Figure 2.28 illustrates primal central paths for two linear optimization problems.

Under mild conditions, $x^\star = \lim_{t \to 0} x(t)$ is a global minimal point of P [37]. However, an obstacle to the practical numerical implementation of this approach is that for $t \searrow 0$ the function $B(t, \cdot)$ possesses a strong curvature near bd M, may even appear to be 'numerically kinked' and is thus difficult to minimize. This is not only reflected in unbounded eigenvalues of the Hessian matrix of $B(t, \cdot)$ for $t \searrow 0$, but also manifests itself in the first-order optimality condition: The optimal point $x(t)$ is the unique solution of

$$0 = \nabla_x B(t, x) = \nabla_x \left(f(x) - t \sum_{i \in I} \log(-g_i(x)) \right)$$

$$= \nabla f(x) - t \sum_{i \in I} \frac{1}{-g_i(x)} (-\nabla g_i(x))$$

$$= \nabla f(x) + \sum_{i \in I} \left(-\frac{t}{g_i(x)} \right) \nabla g_i(x).$$

For $x(t) \to x^* \in \text{bd } M$ there is at least one $i \in I$ with $g_i(x(t)) \to 0$, so the limiting process $\lim_{t \to 0}(-t)/g_i(x(t))$ is 'of type 0/0', and one must expect that the optimality condition for t near zero is numerically unstable.

As a way out, one makes use of the similarity of the above optimality condition to the KKT conditions and sets for all $t > 0$

$$\lambda_i(t) := -\frac{t}{g_i(x(t))}, \quad i \in I.$$

Because of $x(t) \in M_<$ it holds $g_i(x(t)) < 0$ for all $i \in I$, and due to $t > 0$ it follows $\lambda_i(t) > 0$. So $x(t)$ is a minimal point of $B(t, \cdot)$ if and only if some $\lambda(t)$ exists, so that $(x(t), \lambda(t))$ solves the following system of equalities and inequalities:

$$\nabla f(x) + \sum_{i \in I} \lambda_i \nabla g_i(x) = 0,$$

$$\lambda_i g_i(x) = -t,$$

$$\lambda_i > 0,$$

$$g_i(x) < 0, \quad i \in I.$$

This is just the KKT system of P with the right sides of the complementarity conditions perturbed by the expression $-t$, and with strict instead of nonstrict inequalities. The solution $(x(t), \lambda(t))$ is unique for all $t > 0$, $x(t)$ is a 'primal interior point', and $\lambda(t)$ is a 'dual interior point'. The set

$$C_{PD} = \{(x(t), \lambda(t)) | t \in (0, +\infty)\}$$

is called the *primal-dual central path* of P.

For $t \searrow 0$ the perturbed KKT system transforms into the nominal one, i.e., under convergence $\lim_{t \searrow 0}(x(t), \lambda(t)) = (x^*, \lambda^*)$ the point x^* is a KKT point of P with multiplier λ^* and thus a global minimal point. The decisive advantage of this view compared to the barrier method is that, in this setting, for $t \searrow 0$ no numerical issues occur.

Clever implementations of primal-dual interior point methods solve the auxiliary problems for large t only coarsely, but become increasingly more precise for $t \searrow 0$. Moreover, the methods adaptively determine the adjustment of t. For convexly described C^1-problems this can even happen in such a way that the computational effort for the identification of an ε-minimal feasible point grows at most polynomially in the problem dimension. Since this is particularly true for linear optimization problems, primal-dual interior point methods are superior in this respect to the simplex algorithm, for which worst case examples with exponential computational effort are known. This superiority can also be observed in practice for high-dimensional problems, so that modern software packages do not solve such problems using the simplex algorithm, but with primal-dual interior point methods.

2.8 Algorithms

This is rather surprising since, unlike in the simplex algorithm, the linearity of the defining functions of the optimization problem is not exploited at all, but only their convexity. Details on the design and convergence properties of primal-dual interior point methods can be found, for example, in [12, 23, 31].

Attention is required when using primal-dual interior point methods to solve linear optimization problems with a nonunique optimal point set. Unlike the simplex algorithm, which generates a vertex of the optimal facet, here its analytical center is computed, i.e., a point in its relative topological interior (cf. the minimization of x_2 in Fig. 2.28). If, for example, one solves integer linear problems with a totally unimodular system matrix and integer right-hand sides (such as transportation problems) by merely solving the continuous relaxation, and wants to exploit the fact that then every vertex of the feasible set is automatically purely integer, only the simplex algorithm provides an integer optimal point, but primal-dual interior point methods do *not* necessarily. An example of this situation is given in [30].

While, as mentioned above, many results of smooth convex optimization can be generalized to the nonsmooth case using subdifferentials, this only works in a very limited fashion for the efficiency of primal-dual interior point methods. We briefly discuss this for two classes of convex optimization problems where the feasible set is not described by convex C^1-functions.

In the following, for two symmetric (m, m)-matrices A and B, we write the 'inequality' $A \preceq B$ if $B - A \succeq 0$ holds (i.e., if the matrix $B - A$ is positive semidefinite).

Definition 2.8.6 (Spectrahedron) Let symmetric (m, m)-matrices $A_1, \ldots, A_n$ and B be given. Then the set

$$\left\{ x \in \mathbb{R}^n \,\middle|\, \sum_{i=1}^{n} x_i A_i \preceq B \right\}$$

is called a *spectrahedron*.

Exercise 2.8.7 Show that spectrahedra are convex sets.

Exercise 2.8.8 Show that every convex polyhedron can be written as a spectrahedron.

For a vector $c \in \mathbb{R}^n$ and symmetric (m, m)-matrices $A_1, \ldots, A_n$ and B, the problem

$$SDP: \quad \min_{x} c^\mathsf{T} x \quad \text{s.t.} \quad \sum_{i=1}^{n} x_i A_i \preceq B$$

is called a *semidefinite optimization problem*. Due to Exercise 2.8.7, it is a convex optimization problem, and due to Exercise 2.8.8, semidefinite optimization generalizes linear optimization. For numerous applications of semidefinite optimization problems, including ones from structural optimization, control theory, and multiquadratic optimization, we refer to [38]. Nesterov and Nemirovski were able to show in 1988 in [28] that the polynomial complexity of primal-dual interior point methods can be transferred to some classes of nonsmooth convex optimization problems, including semidefinite optimization.

This is *not* the case, however, for a second class of nonsmooth convex optimization problems, which arises from the following seemingly simple modification of the definition of positive semidefiniteness of a matrix: A symmetric (m,m)-matrix A is called *copositive*, if

$$\forall\, d \geq 0 : \quad d^\mathsf{T} A d \geq 0$$

holds. While all positive semidefinite matrices are also copositive, it is easy to construct copositive matrices that are not positive semidefinite. Analogous to spectrahedra, it can again be shown that for symmetric (m,m)-matrices $A_1, \ldots, A_n$ and B the set of $x \in \mathbb{R}^n$ with copositive matrix $B - \sum_{i=1}^{n} x_i A_i$ is convex. Therefore, for a vector $c \in \mathbb{R}^n$ and symmetric (m,m)-matrices $A_1, \ldots, A_n, B$ the *copositive optimization problem*

$$CP : \quad \min_x c^\mathsf{T} x \quad \text{s.t.} \quad B - \sum_{i=1}^{n} x_i A_i \text{ copositive}$$

is a convex optimization problem. Surprisingly, some NP-hard problems of graph theory can be formulated as CPs (for an overview see, for example, [8]), so the existence of polynomial algorithms for the class CP is not to be expected.

As an example, consider the determination of the *stability number* $\alpha(G)$ of an undirected graph G. It denotes the size of the maximum stable set in G, i.e., the largest subset of the vertices of G, such that any two vertices are not adjacent. The determination of $\alpha(G)$ is NP-hard, and $\alpha(G)$ can be written as the optimal value of the copositive problem

$$\min \lambda \quad \text{s.t.} \quad \lambda(I + A_G) - ee^\mathsf{T} \text{ copositive},$$

where A_G denotes the adjacency matrix of G, I the identity matrix and e the vector $(1, \ldots, 1)^\mathsf{T}$.

This shows that convex optimization problems exist that are not efficiently solvable. So when one speaks of convex optimization being 'easy', one must specify which convex problems are being referred to, such as those with smooth describing functions or SDPs.

Of the methods specifically tailored to convexity discussed in Sect. 2.8, the primal-dual interior point methods are of particular practical relevance in modern

2.8 Algorithms

smooth convex optimization. On the other hand, solution methods for general nonlinear optimization problems [37] nowadays are so advanced that even for convex problems they are often superior to Kelley's cutting plane method and the Frank-Wolfe method. Nevertheless, it is important to know these methods, because their basic ideas appear in adapted form, for example, in (mixed-)integer nonlinear optimization [36] and in the global minimization of nonconvex functions, the content of the next chapter.

As a popular modeling approach for convex optimization problems, we finally mention *Disciplined Convex Programming* [13], whose algorithmic implementation is known as CVX.

Nonconvex Optimization Problems

Contents

3.1	Examples and a Conceptual Algorithm	112
3.2	Convex Relaxation	115
3.3	Interval Arithmetic	120
	3.3.1 Motivation and Applications	121
	3.3.2 Basic Interval Operations	122
	3.3.3 Natural Interval Extension	126
	3.3.4 Dependency Effect	128
	3.3.5 Enclosure Property	129
	3.3.6 Taylor Models	131
	3.3.7 Further Notation	132
3.4	Convex Relaxation by the alphaBB Method	133
3.5	Uniformly Refined Tessellations	146
3.6	Branch-and-Bound for Box-Constrained Problems	154
3.7	Branch-and-Bound for Convexly Constrained Problems	162
3.8	Branch-and-Bound for Nonconvex Problems	163
3.9	Lipschitz Properties	168
	3.9.1 Properties of Lipschitz Continuous Functions	170
	3.9.2 Direct Application to Algorithm 3.5	174
	3.9.3 A Variation of Algorithm 3.5	176

In practice, optimization problems are often nonconvex. Simple examples are the projection problem (Example 1.1.1) with nonconvex set M, the problem of cluster analysis (Example 1.1.8) as well as the examples in Sect. 3.1. The common methods to identify global optimal points and values for such problems are based on branch-and-bound ideas. This chapter discusses the αBB method [1, 2] as an exemplary branch-and-bound procedure. Central elements of every branch-and-bound procedure are an intelligent subdivision of the feasible set and the calculation of good lower bounds to the objective function on the resulting subsets. One way to efficiently calculate lower bounds is based on the convex relaxation of nonconvex sets and functions, explained in Sect. 3.2 (for other possibilities such as the exploitation of duality statements see e.g. [7]).

The automated convex relaxation by the αBB method from Sect. 3.4 is largely based on the numerical technique of interval arithmetic, which Sect. 3.3 therefore first presents in some detail. In Sect. 3.5 we pursue a simple uniform subdivision strategy for the feasible set, which however proves to be inefficient. This motivates the use of branch-and-bound techniques in Sect. 3.6, initially only for the simplest case of problems with a box-shaped feasible set. In a further step, Sect. 3.7 discusses the necessary modifications for the case of convex feasible sets, before Sect. 3.8 describes a branch-and-bound procedure for general nonconvex problems. The concluding Sect. 3.9 briefly discusses some possibilities to exploit the Lipschitz continuity of the defining functions in addition to or alternatively to their convexity in branch-and-bound procedures.

3.1 Examples and a Conceptual Algorithm

After two simple examples of nonconvex optimization problems, which are even formulated by means of convex functions, this section formulates a first algorithm for the global solution of nonconvex problems, which, however, will prove to be of little practical use.

Example 3.1.1 (Identification of Redundant Inequalities) Let the set $M = \{x \in \mathbb{R}^n | \ g_i(x) \leq 0, \ i \in I\}$ be nonempty and described by convex functions g_i, $i \in I$. A constraint $g_k(x) \leq 0$ is guaranteed to be *redundant* (i.e., the geometry of M does not change if the condition $g_k(x) \leq 0$ is dropped), if the maximal value v_k of

$$R_k : \quad \max_x \ g_k(x) \quad \text{s.t.} \quad x \in M$$

is negative. Due to the *maximization* of the convex objective function g_k, the optimization problem R_k is *not* convex (unless g_k is even linear).

In Fig. 3.1, the set M is described by three linear inequalities as well as

$$g_4(x) = x_1^2 + x_2^2 - 1 \leq 0.$$

Apparently, dropping the constraint $g_4(x) \leq 0$ enlarges M, so g_4 is not redundant.

The points x^1 and x^2 are local maximal points of R_4 with $g_4(x^1) < 0$ and $g_4(x^2) < 0$. If, using local search methods, only x^1 and x^2 are found, but not the boundary points x of M where g_4 is active, one incorrectly concludes $v_4 < 0$, thus redundancy of the constraint $g_4(x) \leq 0$.

3.1 Examples and a Conceptual Algorithm

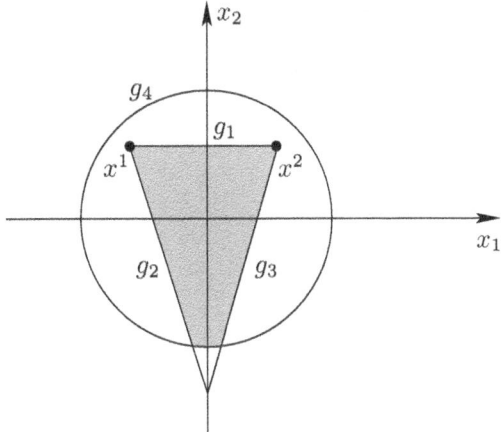

Fig. 3.1 Identification of redundancy as a global optimization problem

Example 3.1.2 (Verification of Convexity) Let $a, b \in \mathbb{R}$, $a < b$, $X = [a, b]$ and $f \in C^2(X, \mathbb{R})$ be given. According to Theorem 2.5.3, f is convex on the (full-dimensional) set X if and only if $f''(x) \geq 0$ holds for all $x \in X$. To verify this criterion, one can try to calculate the minimal value v of

$$K: \quad \min_x f''(x) \quad \text{s.t.} \quad x \in X$$

and test it for $v \geq 0$. However, the optimization problem K cannot be expected to be convex, because even for convex f, f'' does not have to be convex.

In Fig. 3.2, the function $f(x) = x^6 - (3/4)x^5 - 5x^4 + (15/2)x^3 + 15x^2$ and its second derivative are shown on the interval $[-1.5, 1.5]$. The fact that the function f is not convex on this interval can be seen, for example, from the negative values of its second derivative. If one finds only the local minimal point $x^2 = 1$ of the associated problem K with a local search method instead of the global minimal point $x^1 = -1$, one incorrectly concludes $v = f''(x^2) \geq 0$, i.e., the convexity of f.

While in the projection problem and in cluster analysis possibly good local solutions are acceptable, in the decision problems from Example 3.1.1 (g_k redundant or not) and Example 3.1.2 (f convex on X or not) one indeed must determine the global minimal value. However, even in the first case, one needs a measure for what one understands by a 'good local solution'.

We will see that both tasks can be solved in many cases under some effort. For this, we will consider constrained once continuously differentiable minimization problems, i.e., problems of the form

$$P: \quad \min f(x) \quad \text{s.t.} \quad g_i(x) \leq 0, \ i \in I, \ h_j(x) = 0, \ j \in J,$$

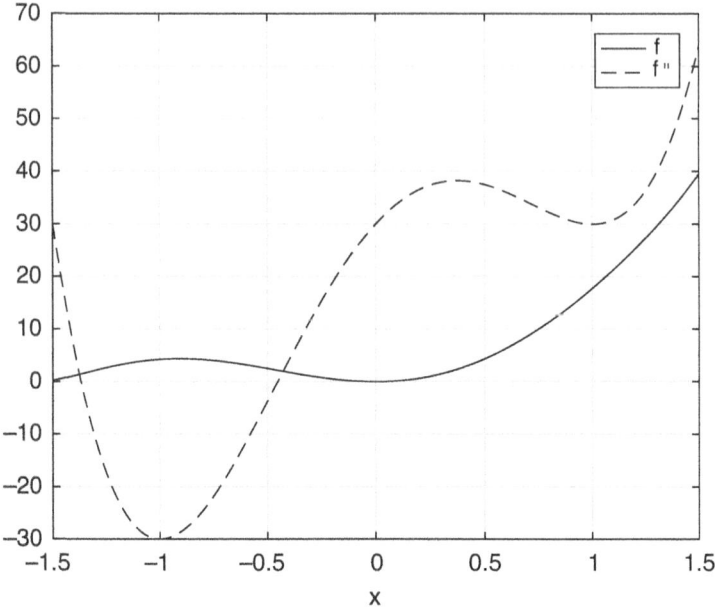

Fig. 3.2 Verification of convexity as a global optimization problem

with continuously differentiable functions f, g_i, $i \in I$, h_j, $j \in J$, which are not necessarily convex. Often we will also assume the solvability of P, which can be checked, for example, by the Weierstrass theorem or its variants from Sec. 1.2.

First, by Theorem 2.7.24, we are able to provide the conceptual Algorithm 3.1 for constrained nonlinear global minimization. It uses the equivalent reformulation of the statement of Theorem 2.7.24 that at every local minimal point $\bar{x}$ of P one of two cases can occur:

- the MFCQ is violated
- the MFCQ is fulfilled and at the same time $\bar{x}$ is a KKT point.

The violation of the MFCQ occurs only in exceptional cases and is therefore also referred to as a *degenerate* case, which explains the notation *DEG* in Algorithm 3.1.

The set $DEG \cup KKT$ is typically not only much smaller than M, but even finite. Then in line 4 the minimization of f on $DEG \cup KKT$ is achieved by comparing finitely many function values.

Example 3.1.3 Figure 3.3 shows a feasible set M described by three C^1-inequalities as well as level lines of a concave-quadratic function f. Because of the cusp of M in x^3, the MFCQ is violated there.

3.2 Convex Relaxation

Fig. 3.3 Candidates for global minimal points

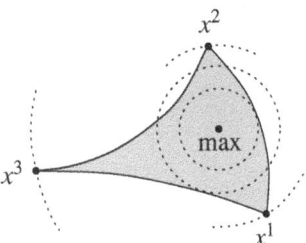

Algorithm 3.1: Conceptual algorithm for constrained nonlinear global minimization

Input: Solvable constrained C^1 optimization problem P
Output: Global minimal point $x^\star$ of P

1 **begin**
2   Determine the set DEG of points in M where the MFCQ is violated.
3   Among the points in M where the MFCQ is fulfilled, determine the set KKT of all KKT points.
4   Determine a minimal point $x^\star$ of f in $DEG \cup KKT$.
5 **end**

For the minimization of f over M, Algorithm 3.1 yields $KKT = \{x^1, x^2\}$ and $DEG = \{x^3\}$. Consequently, in this case $x^\star = x^3$ does not lie in KKT, but in DEG.

Algorithm 3.1 suffers from the following fundamental problem: Unless P is not 'very simple' in a sense explained below, even for finite sets DEG and KKT one cannot be sure whether one has determined *all* of their elements. However, if the sets DEG and KKT are only incompletely calculated, there is a risk that the global minimal points are overlooked. Therefore, Algorithm 3.1 is only applicable when the sets DEG and KKT can be completely determined by, for example, case distinctions.

A considerably more robust solution strategy consists in the exploitation of the fact that at least *smooth convex* problems can be 'easily' solved with current optimization methods (Chap. 2). Therefore, one proceeds analogously to Algorithms 2.1 and 2.2, in which the at the time of publication of the methods still difficult to solve *convex* problems are approximated by easily solvable *linear* problems. We 'lift' this approach by one level and will approximate difficult to solve *nonconvex* problems by easily solvable *convex* problems.

3.2 Convex Relaxation

We establish the connection between nonconvex and convex sets or functions through convex relaxations.

Fig. 3.4 Convex relaxations of a set

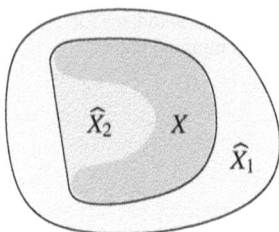

Definition 3.2.1 (Convexly Relaxed Set) Let $X \subseteq \mathbb{R}^n$ be a nonempty set.

(a) Every convex set $\widehat{X} \subseteq \mathbb{R}^n$ with $X \subseteq \widehat{X}$ is called a *convex relaxation* of X.
(b) The intersection of all convex relaxations of X,

$$\widehat{\widehat{X}} := \bigcap \{\widehat{X} \mid \widehat{X} \supseteq X, \ \widehat{X} \text{ convex}\},$$

is called the *convex hull* of X.

In Fig. 3.4, $\widehat{X}_1$ and $\widehat{X}_2$ are convex relaxations of X, and $\widehat{X}_2 = \widehat{\widehat{X}}$ holds.

The convex hull is the smallest possible convex relaxation of X. Due to the convexity of $\widehat{X} = \mathbb{R}^n$ and $X \subseteq \widehat{X}$, it exists for all X. Moreover, it is uniquely determined.

Definition 3.2.2 (Convexly Relaxed Function) Let a nonempty convex set $X \subseteq \mathbb{R}^n$ and a function $f : X \to \mathbb{R}$ be given.

(a) Each function $\widehat{f}$ that is convex on X with

$$\forall x \in X : \widehat{f}(x) \leq f(x)$$

is called a *convex relaxation* of f on X.
(b) A convex relaxation $\widehat{\widehat{f}}$ of f on X, which fulfills

$$\forall x \in X : \widehat{f}(x) \leq \widehat{\widehat{f}}(x)$$

for all other convex relaxations $\widehat{f}$ of f on X, is called *convex hull function* of f on X.

3.2 Convex Relaxation

Fig. 3.5 Convex relaxations of a function

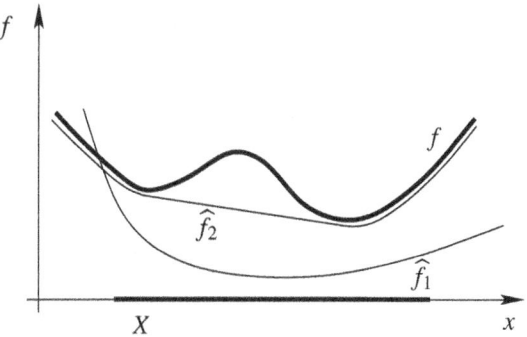

In Fig. 3.5, $\widehat{f}_1$ and $\widehat{f}_2$ are convex relaxations of f on X, and $\widehat{f}_2 = \widehat{\widehat{f}}$ holds. Unlike sets, not every function f on every set X possesses a convex relaxation. For example, $f(x) = -x^2$ has *no* convex relaxation on $X = \mathbb{R}$.

The following theorem shows that convex relaxations of a functionally described set can be obtained using convex relaxations of the describing functions. However, the analogous result does not hold for the convex *hull* of such a set.

Theorem 3.2.3 *Let $X \subseteq \mathbb{R}^n$ be a nonempty convex set (e.g. $X = \mathbb{R}^n$), let $g_i : X \to \mathbb{R}$, $i \in I$, be functions defined on X, and let $M = \{x \in X | \, g_i(x) \leq 0, \, i \in I\}$.*

(a) *If for each $i \in I$ the function $\widehat{g}_i$ is a convex relaxation of g_i on X, then the set*
$$\widehat{M} = \{x \in X | \, \widehat{g}_i(x) \leq 0, \, i \in I\}$$
is a convex relaxation of M.

(b) *Even if for each $i \in I$ the function $\widehat{\widehat{g}}_i$ is the convex hull function of g_i on X, still*
$$\widehat{\widehat{M}} \neq \left\{ x \in X \, \middle| \, \widehat{\widehat{g}}_i(x) \leq 0, \, i \in I \right\}$$
may hold.

Proof To prove statement a, the convexity of the functions $\widehat{g}_i$, $i \in I$, as well as of the set X imply the convexity of the set $\widehat{M}$. Moreover, for all $x \in M$ and $i \in I$

$$\widehat{g}_i(x) \leq g_i(x) \leq 0$$

holds, i.e., $x \in \widehat{M}$. Therefore $\widehat{M}$ is also a relaxation of M, so overall a convex relaxation.

To prove statement b, consider for example $n = |I| = 1$, $X = [0, 2]$ and $g(x) = 1 - x^2$. Then $M = [1, 2]$, $\widehat{\widehat{g}}(x) = 1 - 2x$ and $\{x \in X|\, \widehat{\widehat{g}}(x) \leq 0\} = [1/2, 2]$, but $\widehat{\widehat{M}} = M = [1, 2]$ hold. □

Theorem 3.2.3a provides the usual method for constructing convex relaxations of sets. Theorem 3.2.3b shows, however, that this method cannot be transferred to the construction of convex hulls.

In the next step, we investigate which statements can be made about the relationship between nonconvex optimization problems and problems in which their objective function and feasible set have each been convexly relaxed.

Definition 3.2.4 (Convexly Relaxed Optimization Problem) For a nonempty convex set $X \subseteq \mathbb{R}^n$, let a function $f : X \to \mathbb{R}$, a set $M \subseteq X$ and the optimization problem

$$P : \quad \min f(x) \quad \text{s.t.} \quad x \in M$$

be given.

(a) Let the function $\widehat{f}$ be a convex relaxation of f on X, and $\widehat{M} \subseteq X$ a convex relaxation of M. Then

$$\widehat{P} : \quad \min \widehat{f}(x) \quad \text{s.t.} \quad x \in \widehat{M}$$

is called a *convex relaxation* of P (on X).

(b) Let the convex hull function $\widehat{\widehat{f}}$ of f on X exist, and let $\widehat{\widehat{M}}$ be the convex hull of M. Then

$$\widehat{\widehat{P}} : \quad \min \widehat{\widehat{f}}(x) \quad \text{s.t.} \quad x \in \widehat{\widehat{M}}$$

is called the *convex hull problem* of P (on X).

Note that, in Definition 3.2.4a, $\widehat{M} \subseteq X$ must be required so that $\widehat{f}$ is defined on $\widehat{M}$, but in Definition 3.2.4b the condition $\widehat{\widehat{M}} \subseteq X$ is automatic, because X itself represents a convex relaxation of M.

3.2 Convex Relaxation

Theorem 3.2.5 *With the assumptions from Definition 3.2.4, let the optimization problems P, $\widehat{P}$ and $\widehat{\widehat{P}}$ appearing in statements a to e be solvable with global minimal values v, $\widehat{v}$ and $\widehat{\widehat{v}}$, respectively.*

(a) *For the minimal value $\widehat{v}$ of each convex relaxation $\widehat{P}$ of P, $\widehat{v} \leq v$ holds.*
(b) *For the minimal value $\widehat{\widehat{v}}$ of the convex hull problem $\widehat{\widehat{P}}$ of P and the minimal value $\widehat{v}$ of each convex relaxation $\widehat{P}$, $\widehat{v} \leq \widehat{\widehat{v}} \leq v$ holds (i.e., $\widehat{\widehat{v}}$ is the best lower bound of v achievable by convex relaxation).*
(c) *In general, $\widehat{\widehat{v}} = v$ does not necessarily hold.*
(d) *For $M = X$, $\widehat{\widehat{v}} = v$ holds.*
(e) *If f is linear, $\widehat{\widehat{v}} = v$ holds.*

Proof The proof of statement a is provided by the chain of inequalities

$$\widehat{v} = \min_{x \in \widehat{M}} \widehat{f}(x) \stackrel{M \subseteq \widehat{M}}{\leq} \min_{x \in M} \widehat{f}(x) \stackrel{M \subseteq X}{\leq} \min_{x \in M} f(x) = v.$$

The proof of statement b is left as an exercise for the reader. To see statement c, consider for example $n = 1$, $X = [-2, 2]$, $M = [-2, -1] \cup [1, 2]$ and $f(x) = x^2$. Then $v = 1$ holds, but due to $\widehat{\widehat{M}} = X$ and $\widehat{\widehat{f}}(x) = x^2$ we obtain $\widehat{\widehat{v}} = 0 < 1 = v$.

To prove statement d, due to statement b, only $v \leq \widehat{\widehat{v}}$ needs to be shown. We note that the constant function $\widehat{f}(x) = v$ is a convex relaxation of f on the convex set $M = X$. For the convex hull function $\widehat{\widehat{f}}$ of f on X, it therefore holds

$$\forall x \in X : v = \widehat{f}(x) \leq \widehat{\widehat{f}}(x)$$

and thus $v \leq \min_{x \in X} \widehat{\widehat{f}}(x)$. The convexity of $M = X$ finally implies $\widehat{\widehat{M}} = M$, so

$$v \leq \min_{x \in M} \widehat{\widehat{f}}(x) = \min_{x \in \widehat{\widehat{M}}} \widehat{\widehat{f}}(x) = \widehat{\widehat{v}}.$$

For the proof of statement e, we set $f(x) = c^\mathsf{T} x + d$ with suitable $c \in \mathbb{R}^n$ and $d \in \mathbb{R}$. Due to statement b, only $v \leq \widehat{\widehat{v}}$ needs to be shown. In fact, by definition of the minimal value v,

$$\forall x \in M : \quad v \leq f(x) = c^\mathsf{T} x + d$$

holds, so that the set $\widehat{M} := \{x \in \mathbb{R}^n \mid v \leq c^\mathsf{T} x + d\}$ forms a convex relaxation of M (at this point, only the *concavity* of f is used). That the convex hull $\widehat{\widehat{M}}$ of M is in particular contained in $\widehat{M}$, can be explicitly written as

$$\forall\, x \in \widehat{M}: \quad v \,\leq\, c^\mathsf{T} x + d \,=\, f(x) \,=\, \widehat{f}(x),$$

where we used the *convexity* of f for the last equality, and from which $v \leq \widehat{v}$ follows.

$\square$

The reason for the lacking identity of $\widehat{v}$ and v in Theorem 3.2.5c lies in the fact that $\widehat{f}$ can attain better values on $\widehat{M} \setminus M$ than f on M. At least this effect can be circumvented by a preliminary epigraph reformulation of P.

Exercise 3.2.6 Show that, under the conditions of Theorem 3.2.5, v coincides with the optimal value of the convex problem

$$\min_{x,\alpha} \alpha \quad \text{s.t.} \quad (x,\alpha) \in \widehat{\text{epi}(f,M)}.$$

Recalling from Theorem 3.2.3b, however, that in general not even a functional representation of $\widehat{M}$ by the convex hull functions $\widehat{g}_i$, $i \in I$, is possible, one cannot expect an easily computable functional representation of the convex hull of the epigraph from Exercise 3.2.6. This result is therefore rather of theoretical nature and rarely practically applicable.

For the same reason, there is no reason to actually determine convex hull problems of nonconvex optimization problems. Instead, from now on we will only work with convex relaxations. For the calculation of lower bounds on nonconvex functions, the relation $\widehat{v} \leq v$ from Theorem 3.2.5a will be essential. The *convexity* of the relaxation $\widehat{P}$ of P plays no role for the validity of this inequality, but for a convexly described C^1-problem $\widehat{P}$, $\widehat{v}$ is efficiently *computable* with the methods from Chap. 2.

3.3 Interval Arithmetic

The next step is to clarify whether and how one can algorithmically construct convex relaxations of functions on sets. For this purpose, we make use of *interval arithmetic*, a numerical technique for enclosing errors in function evaluations. After a motivation of this method in Sect. 3.3.1, Sect. 3.3.2 introduces a generalization of the four basic arithmetic operations on intervals as well as the concept of an elementary function. Section 3.3.3 explains how to generalize compositions of elementary functions with basic arithmetic operations to interval arguments. The dependency effect that occurs in this context is inconvenient, as Sect. 3.3.4 explains, because it often prevents tight error bounds. Nevertheless, interval arithmetic always provides enclosures for errors, as proven in Sect. 3.3.5. An improvement of the basic ideas of interval arithmetic through Taylor models is briefly discussed in

3.3 Interval Arithmetic

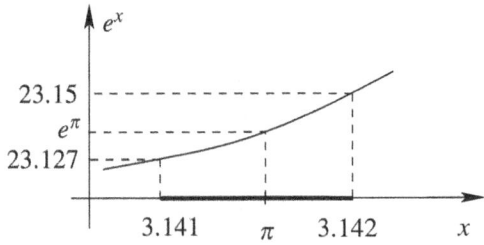

Fig. 3.6 Calculation of e^π

Sect. 3.3.6, before Sect. 3.3.7 provides some notation that we will use in the context of multidimensional intervals.

3.3.1 Motivation and Applications

Interval arithmetic aims at determining quickly computable upper and lower bounds for function values. The basic idea becomes clear using the example of calculating the number e^π: An inaccurate calculation of the form

$$\pi \approx 3.141 \quad \Rightarrow \quad e^\pi \approx e^{3.141} \approx 23.14$$

carries an uncertainty about the result, because it is not clear how far 23.14 lies from the actual value e^π. A better statement is obtained by using lower and upper bounds on the argument π and exploiting the monotonicity of the exponential function to determine lower and upper bounds on the function value e^π (Fig. 3.6):

$$\pi \in [3.141, 3.142]$$

$$e^{3.141} > 23.126, \ e^{3.142} < 23.15 \quad \text{('outward rounding')}$$

$$\Rightarrow e^\pi \in [23.126, 23.15].$$

Therefore the imprecise value calculated above, 23.14, deviates from the actual function value by at most the error $\max\{23.14 - 23.126, 23.15 - 23.14\} = 0.014$.

More generally, for $a \leq b$ and a continuous function $f : [a, b] \to \mathbb{R}$, the image set

$$\text{img}(f, [a, b]) := \{ f(x) | \ x \in [a, b] \}$$

is always a nonempty and compact interval (due to the intermediate value Theorem [18] and the Weierstrass theorem). We avoid the also common notation $f([a, b])$ for img(f, $[a, b]$) here, as it would lead to confusion later in the context of interval arithmetic.

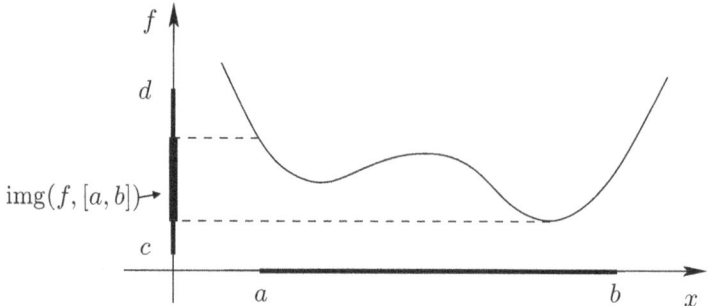

Fig. 3.7 Enclosure of the image set img(f, [a, b])

We are looking for a possibly small interval [c, d] with img(f, [a, b]) $\subseteq$ [c, d], i.e. with

$$\forall x \in [a, b]: \quad c \leq f(x) \leq d$$

(Fig. 3.7).

In view of $c \leq \min_{x \in [a,b]} f(x)$ and $d \geq \max_{x \in [a,b]} f(x)$ the determination of c and d seems to involve two global optimization problems. With methods of interval arithmetic, one can *circumvent* the solution of these problems, but at the price of often only obtaining coarse bounds c and d.

In the numerical calculation of c and d, rounding is often necessary. For safety, one then does not use the usual rounding rule, but 'outward rounding', i.e., c is rounded down, d is rounded up.

In addition to dealing with rounding errors, other applications of interval arithmetic include tolerance analysis in technical and physical systems, the reliable solution of equations (e.g. [16]), and also the computer-assisted proof of Kepler's conjecture [15].

3.3.2 Basic Interval Operations

Interval arithmetic is based on first extending the basic arithmetic operations from numbers to intervals. For this, we introduce the following notation: With

$$\mathbb{IR} = \{[a, b] \mid a, b \in \mathbb{R},\ a \leq b\}$$

we denote the set of all nonempty and compact intervals in $\mathbb{R}$. Elements of $\mathbb{IR}$ are denoted with capital letters, and the interval boundaries with the corresponding lowercase letters as follows:

$$X \in \mathbb{IR} \implies X = \left[\underline{x}, \overline{x}\right].$$

3.3 Interval Arithmetic

For $\underline{x}, \overline{x} \in \mathbb{R}^n$ with $\underline{x} \leq \overline{x}$ the set

$$X = [\underline{x}, \overline{x}] = \{x \in \mathbb{R}^n \mid \underline{x} \leq x \leq \overline{x}\}$$

is a nonempty and compact n-dimensional interval, shortly called *box*. $\mathbb{IR}^n$ denotes the set of all boxes. Alternatively, a box can also be written as $X = [\underline{x}_1, \overline{x}_1] \times \ldots \times [\underline{x}_n, \overline{x}_n]$ (i.e., roughly speaking as a 'vector of intervals' instead of an 'interval of vectors').

In the following we derive 'natural' definitions of basic arithmetic operations on $\mathbb{IR}$. The systematic for this consists in calculating the image sets of the corresponding operations for 'interval inputs'. Formally written, for a (continuous) operation $f : \mathbb{R}^2 \to \mathbb{R}$, $(x, y) \mapsto f(x, y)$ (such as $f(x, y) = x + y$) a corresponding operation $F : \mathbb{IR}^2 \to \mathbb{IR}$ is determined by $(X, Y) \mapsto \text{img}(f, X \times Y)$, where the set $\text{img}(f, X \times Y)$ is explicitly calculated. Since f is continuous and $\mathbb{IR}^2$ only consists of nonempty and compact sets, the intermediate value theorem and the Weierstrass theorem again guarantee that F actually maps to $\mathbb{IR}$, that is, the 'output' of F is always a nonempty and compact interval.

Addition

For $X, Y \in \mathbb{IR}$ the (Minkowski) sum $X + Y$ is defined as the set of all occurring sums of elements from X and Y:

$$X + Y := \{x + y \mid x \in X, \ y \in Y\}.$$

This fits into the above systematics by choosing $f(x, y) = x + y$ and $F(X, Y) = \text{img}(f, X \times Y) = \{x + y \mid (x, y) \in X \times Y\} = X + Y$.

Next, the representation of $X + Y$ must be made explicit: With $X = [\underline{x}, \overline{x}]$ and $Y = [\underline{y}, \overline{y}]$ it holds

$$X + Y = \left[\underline{x} + \underline{y}, \overline{x} + \overline{y}\right],$$

because for all $x \in X$ and $y \in Y$ the addition of the two inequalities $\underline{x} \leq x \leq \overline{x}$ and $\underline{y} \leq y \leq \overline{y}$ results in the new inequalities

$$\underline{x} + \underline{y} \leq x + y \leq \overline{x} + \overline{y},$$

from which $X + Y \subseteq [\underline{x} + \underline{y}, \overline{x} + \overline{y}]$ follows. Since also the points $\underline{x} + \underline{y}$ and $\overline{x} + \overline{y}$ lie in $X + Y$ and, being an interval, $X + Y$ is a convex set, also $X + Y \supseteq [\underline{x} + \underline{y}, \overline{x} + \overline{y}]$ holds, thus proving the assertion. We therefore obtain the definition

$$\left[\underline{x}, \overline{x}\right] + \left[\underline{y}, \overline{y}\right] := \left[\underline{x} + \underline{y}, \overline{x} + \overline{y}\right].$$

It is easy to see that the above formula also applies for $X, Y \in \mathbb{IR}^n$, so that we have defined the addition of boxes of any dimension as well.

Addition with a Vector

For $x \in \mathbb{R}^n$ and $Y = [\underline{y}, \overline{y}] \in \mathbb{IR}^n$ it suggests itself to define the addition

$$x + [\underline{y}, \overline{y}] := [x + \underline{y}, x + \overline{y}].$$

Indeed, the above systematics yield this result. An alternative approach would be to *identify* the vector $x \in \mathbb{R}^n$ with the singleton box $[x, x] \in \mathbb{IR}^n$ and then apply the above rule for two intervals. For the sake of clarity, however, in the following we will consistently *refrain* from this identification of a vector with a singleton box.

The definitions of the remaining arithmetic operations are derived in the same systematic way.

Subtraction

For $X, Y \in \mathbb{IR}^n$ we set

$$-X := \{-x |\, x \in X\} = [-\overline{x}, -\underline{x}]$$

and

$$X - Y := X + (-Y) = [\underline{x} - \overline{y}, \overline{x} - \underline{y}].$$

Multiplication

For $X, Y \in \mathbb{IR}$ we set

$$X \cdot Y := \{x\, y|\, x \in X,\ y \in Y\} = \Box \{\underline{x}\,\underline{y},\ \underline{x}\,\overline{y},\ \overline{x}\,\underline{y},\ \overline{x}\,\overline{y}\},$$

where for a bounded set of real numbers $A \subseteq \mathbb{R}$ we introduce the *interval hull*

$$\Box A := [\inf A,\ \sup A]$$

of A, as common in the literature on interval arithmetic. For the *finite* sets $A \subseteq \mathbb{R}$ that we only need in our applications, $\Box A$ coincides with the convex hull of A.

Due to the lack of a common multiplication rule for vectors $x, y \in \mathbb{R}^n$ that yields a result in $\mathbb{R}^n$, we neither define a multiplication of boxes $X, Y \in \mathbb{IR}^n$. Of course, an interval analogue of the inner product of two vectors may be defined, but it would not be referred to as an arithmetic operation.

3.3 Interval Arithmetic

Multiplication with a Scalar

For $x \in \mathbb{R}$ with $x \geq 0$ and $Y = [\underline{y}, \overline{y}] \in \mathbb{IR}^n$ we define

$$x \cdot [\underline{y}, \overline{y}] := [x\underline{y}, x\overline{y}]$$

and for $x < 0$

$$x \cdot [\underline{y}, \overline{y}] := (-x) \cdot (-Y) = [x\overline{y}, x\underline{y}].$$

Division

For $X \in \mathbb{IR}$ with $0 \notin X$ we set

$$\frac{1}{X} := \left\{ \frac{1}{x} \middle| x \in X \right\} = \left[\frac{1}{\overline{x}}, \frac{1}{\underline{x}}\right]$$

and for $X, Y \in \mathbb{IR}$ with $0 \notin Y$

$$\frac{X}{Y} := X \cdot \left(\frac{1}{Y}\right) = \square \left\{ \frac{\underline{x}}{\underline{y}}, \frac{\underline{x}}{\overline{y}}, \frac{\overline{x}}{\overline{y}}, \frac{\overline{x}}{\underline{y}} \right\}.$$

Now that the basic arithmetic operations for intervals are defined, we first note that *not all arithmetic rules from $\mathbb{R}$ can be transferred to intervals*. For example, for $X \in \mathbb{IR}$ in general neither $X - X = [0, 0]$ nor $X/X = [1, 1]$ holds, but this is only true for $\underline{x} = \overline{x}$. Also, $X \cdot X$ does not yield the image of X under the function $f(x) = x^2$ for every $X \in \mathbb{IR}$. This is due to the *dependency effect*, which is explained in Sect. 3.3.4 and always occurs when the *same* interval variable appears *multiple times* in function expressions.

With the help of the basic arithmetic operations, *rational interval-valued functions* can be defined, i.e., functions $F : \mathbb{IR}^n \to \mathbb{IR}$, in whose function expression only the basic interval arithmetic operations appear.

Example 3.3.1 For $F : \mathbb{IR}^2 \to \mathbb{IR}$, $F(X_1, X_2) = ([1, 2]X_1 + [0, 1])X_2$ and $X_1 = X_2 = [0, 1]$ we calculate $F(X_1, X_2)$. This is done step by step as follows:

$$V_1 = [1, 2]X_1 = [1, 2][0, 1] = \square\{0, 1, 2\} = [0, 2],$$
$$V_2 = V_1 + [0, 1] = [0, 2] + [0, 1] = [0, 3],$$
$$F(X_1, X_2) = V_2 X_2 = [0, 3][0, 1] = [0, 3].$$

In addition to the basic arithmetic operations, elementary functions f (e.g., functions predefined in the software environment available to the user) can also be naturally extended to interval-valued functions according to the above scheme, again by explicitly calculating $F(X) := \text{img}(f, X)$. For example, the monotonicity of the functions exp and log yields

$$\text{EXP}(X) := [\exp(\underline{x}), \exp(\overline{x})]$$

for $X \in \mathbb{IR}$ and

$$\text{LOG}(X) := [\log(\underline{x}), \log(\overline{x})]$$

for $X \in \mathbb{IR}$ with $\underline{x} > 0$. Due to the lack of monotonicity of the function sin, the calculation of $\text{SIN}(X)$ for $X \in \mathbb{IR}$ as img(sin, X) is less simple, but still possible through appropriate case distinctions.

Exercise 3.3.2 Calculate the set img(sin, X) for all $X \in \mathbb{IR}$.

In the following, we will occasionally only need the explicit lower or upper boundary of an interval $F(X)$ and write

$$F(X) = [\underline{F}(X), \overline{F}(X)]$$

for this. For example, one obtains $\underline{\text{EXP}}([\underline{x}, \overline{x}]) = \exp(\underline{x})$ and $\overline{\text{EXP}}([\underline{x}, \overline{x}]) = \exp(\overline{x})$.

3.3.3 Natural Interval Extension

For the following definition, recall that for $f : \mathbb{R}^k \to \mathbb{R}$ and $g : \mathbb{R}^n \to \mathbb{R}^k$ the function $f \circ g : \mathbb{R}^n \to \mathbb{R}$, $x \mapsto (f \circ g)(x) := f(g(x))$ is called *composition* of f and g.

> **Definition 3.3.3 (Factorizable Function)** A function $f : \mathbb{R}^n \to \mathbb{R}$ is called *factorizable*, if the function expression of f can be broken down into a finite number of elementary operations, consisting of the basic arithmetic operations, elementary functions, and compositions.

Example 3.3.4 The function $f(x) = (\sin(e^{x_1 + 3x_2}) + 5)/(\|x\|_2 + 1)$ is factorizable, the function $f(x) = \sum_{k=0}^{\infty} x^k/(2k)!$ is *not*, and also solutions of differential equations are often *not*.

Definition 3.3.3 implies that the techniques of interval arithmetic base on the *representation* of functions, rather than on their expression independent properties. For example, the functions $f_1(x) = \exp(x) - 1$ and $f_2(x) = \sum_{k=1}^{\infty} x^k/k!$ are identical, but only the representation f_1 forms a factorization. Moreover,

3.3 Interval Arithmetic

factorizability can depend rather individually on which elementary functions a user encounters in the underlying software environment.

> **Definition 3.3.5 (Interval Extension)**
>
> (a) For $f : \mathbb{R}^n \to \mathbb{R}$, $F : \mathbb{IR}^n \to \mathbb{IR}$ is called *interval extension of f*, if $\forall x \in \mathbb{R}^n : F([x, x]) = [f(x), f(x)]$ holds.
> (b) For a factorizable function $f : \mathbb{R}^n \to \mathbb{R}$ with given factorization $F : \mathbb{IR}^n \to \mathbb{IR}$, $F(X_1, \ldots, X_n) := f(X_1, \ldots, X_n)$ is called *natural interval extension of f*.

In part b of this definition, $f(X_1, \ldots, X_n)$ stands for the function expression which arises when in the present factorization of f each occurrence of a variable x_i is replaced by X_i and then all occurring elementary operations are interpreted as interval-valued. Crucial for the construction of interval arithmetic is that this does *not* mean the set $\mathrm{img}(f, X_1 \times \ldots \times X_n)$. As Example 3.3.7 will show, $f(X_1, \ldots, X_n)$ and $\mathrm{img}(f, X_1 \times \ldots \times X_n)$ generally do not coincide. The fact that for $n = 1$ in particular $f(X)$ and $\mathrm{img}(f, X)$ do not need to coincide is the reason for our notation of the image of X under f.

Exercise 3.3.9 will make sure that the natural interval extension from Definition 3.3.5b is an interval extension in the sense of Definition 3.3.5a. Finally, if a function f is not defined on all of $\mathbb{R}^n$, like $f = \log$ for $n = 1$ or the division $f(x, y) = x/y$ for $n = 2$, then the domain of the interval extension from Definition 3.3.5 must also be constrained to a suitable subset of $\mathbb{IR}^n$.

Example 3.3.6 The functions EXP and LOG defined above are interval extensions of the elementary functions exp and log. The same applies to the basic interval operations, e.g. $F([\underline{x}, \overline{x}], [\underline{y}, \overline{y}]) = [\underline{x} + \underline{y}, \overline{x} + \overline{y}]$ is an interval extension of $f(x, y) = x + y$. The function $F : \mathbb{IR}^2 \to \mathbb{IR}$ from Example 3.3.1 however, cannot be the interval extension of a function $f : \mathbb{R}^2 \to \mathbb{R}$.

Example 3.3.7 The graph of $f(x) = x - x^2$ is shown in Fig. 3.8. It is easy to see that this function satisfies $\mathrm{img}(f, [0, 1]) = [0, 1/4]$.

Since f can be factorized as $f(x) = x - x \cdot x$, f has the natural interval extension $F(X) = X - X \cdot X$. Substituting $X = [0, 1]$ yields

$$F([0, 1]) = [0, 1] - [0, 1][0, 1] = [0, 1] - [0, 1] = [-1, 1],$$

so that $F([0, 1])$ does *not* coincide with $\mathrm{img}(f, [0, 1])$.

It gets even worse: The function f can alternatively be factorized by $\widetilde{f}(x) = x(1 - x)$, but despite $f = \widetilde{f}$, the natural interval extension $\widetilde{F}(X) = X(1 - X)$ of $\widetilde{f}$ does not coincide with the natural interval extension F of f, because for $X = [0, 1]$

Fig. 3.8 Image set of f on $[0, 1]$

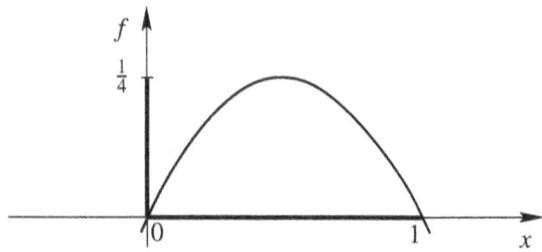

we obtain

$$\widetilde{F}([0, 1]) = [0, 1](1 - [0, 1]) = [0, 1][0, 1] = [0, 1] \neq [-1, 1] = F([0, 1]).$$

3.3.4 Dependency Effect

Example 3.3.7 shows that not only the already discussed applicability, but also the *results* of interval arithmetic can strongly depend on the representation of the functions considered, rather than just on their expression independent properties. The reason for this lies in the *dependency effect*, which one always has to reckon with when an interval variable appears more than once in a function expression. It is caused by the fact that a variable appearing multiple times is treated in the same manner as several independent variables.

For example, $X(1 - X)$ is interpreted in the same way as $X(1 - Y)$ with the additional condition $X = Y$. The latter dependency of X and Y is however ignored by interval arithmetic. This leads to the fact that when considering the two factors x and $1 - x$ dependencies like

$$x = 1 \Rightarrow 1 - x = 0 \quad \text{and} \quad x = \frac{1}{2} \Rightarrow 1 - x = \frac{1}{2}$$

are ignored, so that in the calculation of $X(1 - X)$ also products like $1 \cdot 1$ appear, which do not occur in the evaluation of $x(1 - x)$.

Similarly, $X - X$ is treated like $X - Y$ with $X = Y$ and X/X like X/Y with $X = Y$, which is why $X - X = [0, 0]$ or $X/X = [1, 1]$ are wrong in general. Also the product $X \cdot X$ does not always yield the image of X under $f(x) = x^2$ (e.g. for $X = [-1, 1]$), although we have defined the product of X and Y as the exact image of the product operator. The latter is again only true for two independent inputs X and Y, while $X \cdot X$ is treated like $X \cdot Y$ with $X = Y$.

As a rule of thumb one may formulate: The more often a variable appears, the larger the interval calculated by interval arithmetic becomes. Example 3.3.7 shows, for instance, that $X - X \cdot X$ (X appears three times) yields a considerably larger interval for $X = [0, 1]$ than $X(1 - X)$ (X only appears twice).

3.3 Interval Arithmetic

Frequently occurring functions that suffer from the dependency effect can be better handled by interval arithmetic by also considering them as elementary functions, for example all monomials x^k, $k \in \mathbb{N}$. Indeed, $f(x) = x \cdot x$ has the interval extension $F(X) = X \cdot X$ with $F([-1, 1]) = [-1, 1]$, while considering sqr$(x) = x^2$ as an elementary function allows the interval extension

$$\text{SQR}([\underline{x}, \overline{x}]) = \begin{cases} [\min\{\underline{x}^2, \overline{x}^2\}, \max\{\underline{x}^2, \overline{x}^2\}], & \text{if } 0 \notin [\underline{x}, \overline{x}] \\ [0, \max\{\underline{x}^2, \overline{x}^2\}], & \text{if } 0 \in [\underline{x}, \overline{x}] \end{cases}$$

with $\text{SQR}([-1, 1]) = [0, 1]$. Also functions like norm$(x) = \|x\|_2$ are often considered elementary.

Exercise 3.3.8 Verify that, with the given function SQR, for all $X \in \mathbb{IR}$ the relationship $\text{SQR}(X) = \text{img}(\text{sqr}, X)$ holds.

3.3.5 Enclosure Property

Despite these rather 'unpleasant' properties of the natural interval extension, it fulfills our main purpose of delivering a *superset* of img(f, X) as a result of $F(X)$, thus providing a guaranteed lower bound on the minimal value and a guaranteed upper bound on the maximal value of f on X. This is also illustrated by Example 3.3.7:

$$\left[0, \tfrac{1}{4}\right] = \text{img}(f, [0, 1]) \subseteq \begin{cases} F([0, 1]) = [-1, 1] \\ \widetilde{F}([0, 1]) = [0, 1]. \end{cases}$$

In the following we will derive that this property is always true. First, however, we need to specify the interval extension of the composition of two functions more precisely.

A *vector-valued* function $f : \mathbb{R}^n \to \mathbb{R}^m$ is called factorizable if each component f_j, $j = 1, \ldots, m$, is factorizable as a function from $\mathbb{R}^n$ to $\mathbb{R}$, and its natural interval extension is then defined componentwise as

$$F(X_1, \ldots, X_n) := \begin{pmatrix} F_1(X_1, \ldots, X_n) \\ \vdots \\ F_m(X_1, \ldots, X_n) \end{pmatrix}$$

with the natural interval extensions F_j of the factorized functions f_j, $j = 1, \ldots, m$.

If $f : \mathbb{R}^k \to \mathbb{R}$ and $g : \mathbb{R}^n \to \mathbb{R}^k$ are factorizable functions with natural interval extensions $F : \mathbb{IR}^k \to \mathbb{IR}$ and $G : \mathbb{IR}^n \to \mathbb{IR}^k$, then one defines, quite naturally,

$$(F \circ G)(X) := F(G(X))$$

as the natural interval extension of $f \circ g$. This is merely the formalization of the fact that $F(G(X))$ is used as the interval extension of $f(g(x))$.

Exercise 3.3.9 Show that for every factorizable function $f : \mathbb{R}^n \to \mathbb{R}$ and each of its factorizations, its natural interval extension introduced in Definition 3.3.5b is not only *called* an interval extension, but actually *is* an interval extension of f in the sense of Definition 3.3.5a.

Definition 3.3.10 (Monotone Interval Extension) An interval-valued function $F : \mathbb{IR}^n \to \mathbb{IR}$ is called *monotone* (or *inclusion isotonic*), if

$$\forall X, Y \in \mathbb{IR}^n \text{ with } X \subseteq Y : \quad F(X) \subseteq F(Y)$$

holds.

Theorem 3.3.11 *The interval versions of the basic arithmetic operations, of elementary functions, and of compositions of functions are monotone.*

Proof For each basic arithmetic operation and each elementary function f, their interval versions satisfy $f(X) = \text{img}(f, X)$ for corresponding sets X. Since $\text{img}(f, X) \subseteq \text{img}(f, Y)$ holds for all $X \subseteq Y$, their monotonicity follows from this.

For functions f and g that do not contain compositions themselves, the monotonicity of $F \circ G$ is clear. Since a composition cannot appear as an 'innermost' operation in a function expression, a recursive argument yields the monotonicity of each composition $F \circ G$. □

Theorem 3.3.11 implies the following result.

Corollary 3.3.12 *For every factorizable function $f : \mathbb{R}^n \to \mathbb{R}$ and each of its factorizations, the natural interval extension $F : \mathbb{IR}^n \to \mathbb{IR}$ is monotone.*

With this, we are able to prove the desired enclosure property.

3.3 Interval Arithmetic

Theorem 3.3.13 *For every factorizable function $f : \mathbb{R}^n \to \mathbb{R}$, for each of its factorizations and corresponding natural interval extensions $F : \mathbb{IR}^n \to \mathbb{IR}$, and for all $X \in \mathbb{IR}^n$*

$$\text{img}(f, X) \subseteq F(X)$$

holds.

Proof For all $x \in X$ we have

$$[f(x), f(x)] = F([x, x]) \stackrel{F \text{ monotone}}{\subseteq} F(X)$$

and thus $f(x) \in F(X)$ and $\text{img}(f, X) \subseteq F(X)$. □

Theorem 3.3.13 means that the natural interval extension provides valid bounds for $\text{img}(f, X)$. As seen, however, those may be coarse due to the dependency effect.

3.3.6 Taylor Models

Improved bounds can be achieved with *Taylor models*, the idea of which is briefly explained in the following. For continuously differentiable $f : M \to \mathbb{R}$ with some convex set $M \subseteq \mathbb{R}^n$, the mean value theorem yields

$$\forall x, \tilde{x} \in M : \quad f(x) = f(\tilde{x}) + \langle \nabla f(y), x - \tilde{x} \rangle$$

with some y on the connecting line segment of x and $\tilde{x}$. Not much is known about y, but at least y must also lie in M, because it lies in the convex hull of the elements x and $\tilde{x}$ of M, and M is convex. Since with the abbreviation $g := \nabla f$

$$\forall x, \tilde{x} \in M : \quad f(x) = f(\tilde{x}) + \sum_{i=1}^{n} g_i(y)(x_i - \tilde{x}_i)$$

holds, it follows

$$\forall x, \tilde{x} \in M : \quad f(x) \in f(\tilde{x}) + \sum_{i=1}^{n} \text{img}(g_i, M)(x_i - \tilde{x}_i).$$

In particular, if one chooses $M \in \mathbb{IR}^n$ and moreover $g = \nabla f$ is factorizable with natural interval extension G, then one does not need to explicitly calculate the sets

$\mathrm{img}(g_i, M)$, $i = 1, \ldots, n$, but, due to $\mathrm{img}(g_i, M) \subseteq G_i(M)$, can conclude

$$\forall x, \tilde{x} \in M: \quad f(x) \in f(\tilde{x}) + \sum_{i=1}^{n} G_i(M)(x_i - \tilde{x}_i).$$

Thus, for all $X \in \mathbb{R}^n$ with $X \subseteq M$ and each $\tilde{x} \in M$

$$\mathrm{img}(f, X) \subseteq f(\tilde{x}) + \sum_{i=1}^{n} G_i(M)([\underline{x}_i, \overline{x}_i] - \tilde{x}_i)$$

holds. For $X \subseteq M$ the expression $F(X) := f(\tilde{x}) + \sum_{i=1}^{n} G_i(M)([\underline{x}_i, \overline{x}_i] - \tilde{x}_i)$ is not an interval extension of f, but still provides bounds for $\mathrm{img}(f, X)$, and in particular it follows

$$\mathrm{img}(f, M) \subseteq F(M) = f(\tilde{x}) + \sum_{i=1}^{n} G_i(M)([\underline{m}_i, \overline{m}_i] - \tilde{x}_i).$$

As $\tilde{x}$ one may choose, e.g., the midpoint of M, so $\tilde{x} = (\underline{m} + \overline{m})/2$.

Example 3.3.14 For $f(x) = x(1 - x)$ and $M = [0, 1]$ we have $g(x) = f'(x) = 1 - 2x$ and thus $G([0, 1]) = 1 - 2[0, 1] = [-1, 1]$. With $\tilde{x} = 1/2$ we obtain

$$\mathrm{img}(f, [0, 1]) \subseteq F([0, 1]) := f\left(\tfrac{1}{2}\right) + G([0, 1])\left([0, 1] - \tfrac{1}{2}\right)$$

$$= \tfrac{1}{4} + [-1, 1]\left([0, 1] - \tfrac{1}{2}\right) = \tfrac{1}{4} + [-1, 1]\left[-\tfrac{1}{2}, \tfrac{1}{2}\right]$$

$$= \tfrac{1}{4} + \left[-\tfrac{1}{2}, \tfrac{1}{2}\right] = \left[-\tfrac{1}{4}, \tfrac{3}{4}\right].$$

Compare this with the corresponding results from Example 3.3.7.

This idea of a Taylor model can be generalized analogously to Taylor expansions of higher order [16].

3.3.7 Further Notation

For later use, we introduce the following notations for a box $X = [\underline{x}, \overline{x}] \in \mathbb{R}^n$:

- $m(X) = (\underline{x} + \overline{x})/2$ is the *box center*.
- $w(X) = \|\overline{x} - \underline{x}\|_2$ is the *box width*.

Note that $w(X)$ denotes the length of the box diagonal and thus is a measure for the box size. In the literature on interval arithmetic one also finds other choices of the norm for the definition of $w(X)$. For example, the ℓ_∞-norm corresponds to the largest edge length of X, and the ℓ_1-norm provides the sum of the edge lengths along the n coordinate directions, a multiple of which (depending on the dimension) can be considered as the 'perimeter' of X, i.e., the sum of all edge lengths. In the case of $n = 1$, for each of the three mentioned norms the box width collapses to $w(X) = \overline{x} - \underline{x}$.

Exercise 3.3.15 For a set $M \subseteq \mathbb{R}^n$, $\sup_{x,y \in M} \|x - y\|_2$ denotes the *diameter* of M. Show that the diameter of a box X coincides with its box width.

From a numerical point of view, especially when dealing with rounding errors in the evaluation of f at some $x \in \mathbb{R}^n$ by interval arithmetic, a crucial next question would be whether and how quickly the enclosing intervals $F(X)$ for $\mathrm{img}(f, X)$ become small when the enclosing boxes X for x shrink with $w(X) \to 0$. Here, Taylor models are superior to the natural interval extension. However, since this question will not arise for our application in global optimization, we refer to, e.g., [29] for further details.

3.4 Convex Relaxation by the alphaBB Method

With the help of interval arithmetic, convex relaxations of functions can be constructed in various ways. We focus here on the αBB method [1, 2]. For further details on this method, we refer to [10].

Since the techniques of interval arithmetic fit well with box-shaped sets, we first look for a convex relaxation of a function $f \in C^2(X, \mathbb{R})$ on a box $X \in \mathbb{IR}^n$, i.e., for a convex function $\widehat{f} : X \to \mathbb{R}$ with $\widehat{f}(x) \leq f(x)$ for all $x \in X$ (Definition 3.2.2a). The twice continuous differentiability of f on X allows us to use the C^2-characterization of convexity for the construction of $\widehat{f}$. The full dimensionality of the box X is both desirable and not restrictive, which is why from now on we assume boxes $X = [\underline{x}, \overline{x}]$ with $\underline{x} < \overline{x}$.

The basic idea of the αBB method is to first construct a simple strongly convex function ψ that is nonpositive on X. Then one adds a sufficiently large multiple $\alpha \psi$ of ψ to f, i.e., one sets

$$\widehat{f}_\alpha(x) := f(x) + \alpha \psi(x)$$

with sufficiently large $\alpha \geq 0$ to compensate for 'nonconvexities' of f on X.

With the representation $X = [\underline{x}, \overline{x}]$, a simple strongly convex and nonpositive function on X is given by

$$\psi(x) := \frac{1}{2}(\underline{x} - x)^\mathsf{T}(\overline{x} - x),$$

because for all $x \in X$

$$\underline{x}_i \leq x_i \leq \overline{x}_i, \quad i = 1, \ldots, n,$$

implies

$$\psi(x) = \frac{1}{2} \sum_{i=1}^{n} \underbrace{(\underline{x}_i - x_i)}_{\leq 0} \underbrace{(\overline{x}_i - x_i)}_{\geq 0} \leq 0,$$

and, furthermore, at all $x \in \mathbb{R}^n$ the Hessian matrix of ψ is

$$D^2 \psi(x) = \begin{pmatrix} 1 & & 0 \\ & \ddots & \\ 0 & & 1 \end{pmatrix} = I.$$

Thus we obtain $\lambda_{\min}(D^2\psi(x)) = 1$, so that ψ is strongly convex on $\mathbb{R}^n$. We also note that ψ vanishes at every vertex point of X, because for a vertex y of X in each component $i \in \{1, \ldots, n\}$ either $y_i = \underline{x}_i$ or $y_i = \overline{x}_i$ holds. Thus, each summand of $\psi(y)$ vanishes, so $\psi(y) = 0$. Figures 3.9 and 3.10 show the shape of ψ in the cases $n = 1$ and $X = [1, 2]$ as well as $n = 2$ and $X = [1, 2] \times [1, 3]$, respectively.

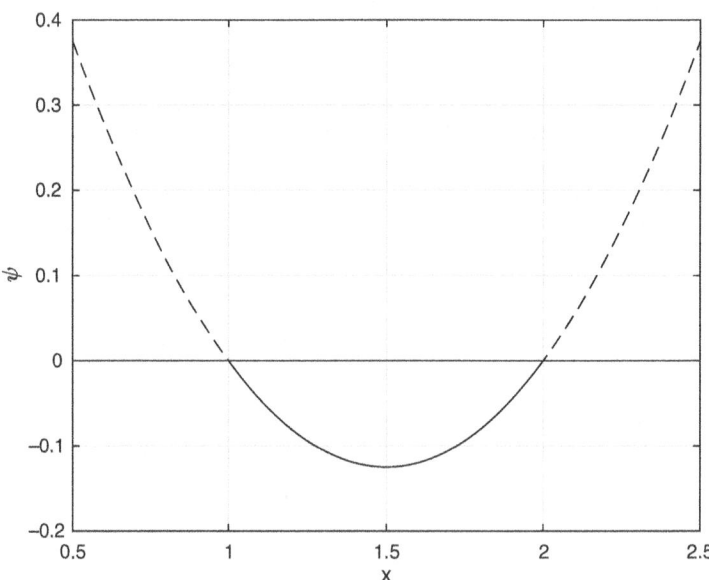

Fig. 3.9 ψ for $n = 1$

3.4 Convex Relaxation by the alphaBB Method

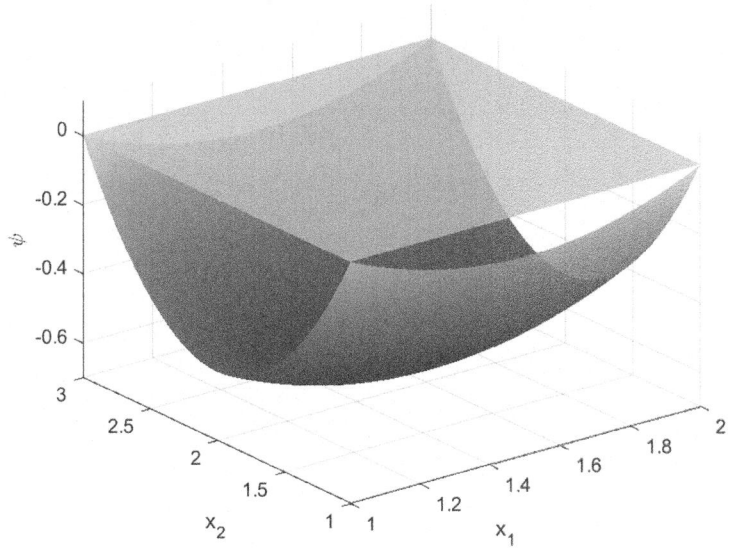

Fig. 3.10 ψ for $n = 2$

Exercise 3.4.1 Show that the unique minimal point of the strongly convex function

$$\psi(x) = \frac{1}{2}(\underline{x} - x)^\mathsf{T}(\overline{x} - x)$$

is the midpoint $m(X) = (\underline{x} + \overline{x})/2$ of the box X with minimal value

$$\min_{x \in \mathbb{R}^n} \psi(x) = \min_{x \in X} \psi(x) = \psi(m(X)) = -\frac{1}{8} w(X)^2, \tag{3.1}$$

where $w(X) = \|\overline{x} - \underline{x}\|_2$ denotes the box width.

So we set

$$\widehat{f_\alpha}(x) = f(x) + \alpha \psi(x) = f(x) + \frac{\alpha}{2}(\underline{x} - x)^\mathsf{T}(\overline{x} - x)$$

with a parameter α to be determined. Without further assumptions, we can already show the following results.

Lemma 3.4.2

(a) For all $x \in X$ and $\alpha \geq 0$, $\widehat{f_\alpha}(x) \leq f(x)$ holds.
(b) For every vertex point y of X and every $\alpha \in \mathbb{R}$, $\widehat{f_\alpha}(y) = f(y)$ holds.
(c) For all $\alpha \geq 0$, the maximal deviation between f and $\widehat{f_\alpha}$ on X satisfies

$$\max_{x \in X} (f(x) - \widehat{f_\alpha}(x)) = \frac{\alpha}{8} w(X)^2.$$

Proof It holds

$$\forall x \in X,\ \alpha \geq 0: \quad f(x) - \widehat{f_\alpha}(x) = -\underbrace{\alpha}_{\geq 0} \underbrace{\psi(x)}_{\leq 0} \geq 0,$$

thus statement a. Moreover, every vertex point y of X and all $\alpha \in \mathbb{R}$ fulfill

$$f(y) - \widehat{f_\alpha}(y) = -\alpha \underbrace{\psi(y)}_{=0} = 0,$$

which proves statement b. Statement c holds due to

$$\max_{x \in X} (f(x) - \widehat{f_\alpha}(x)) = -\alpha \min_{x \in X} \psi(x) \stackrel{(3.1)}{=} \frac{\alpha}{8} w(X)^2$$

for all $\alpha \geq 0$. □

Lemma 3.4.2c states that the maximal deviation between f and $\widehat{f_\alpha}$ on X only depends on α and the box width $w(X)$ (and, for example, not on f or on other geometric properties of X). It is also noteworthy that Lemma 3.4.2c goes beyond providing an upper bound for the maximal error by even stating its exact value.

According to Lemma 3.4.2a, $\widehat{f_\alpha}$ is a relaxation of f on X for all $\alpha \geq 0$. Next, we choose α so large that also the 'nonconvexities' in f are compensated by the (strongly) convex term $\alpha \psi$. In view of $f \in C^2$ and $\psi \in C^2$, also $\widehat{f_\alpha} = f + \alpha \psi$ is a C^2-function. According to the C^2-characterization of convexity (Theorem 2.5.3), $\widehat{f_\alpha}$ is convex on the (full-dimensional) box X if and only if $D^2 \widehat{f_\alpha}(x) \succeq 0$ holds for all $x \in X$, i.e., all eigenvalues of $D^2 \widehat{f_\alpha}(x)$ are nonnegative for all $x \in X$.

Due to

$$D^2 \widehat{f_\alpha}(x) = D^2 f(x) + \alpha D^2 \psi(x) = D^2 f(x) + \alpha I$$

we have to choose α such that all eigenvalues of $D^2 f(x) + \alpha I$ are nonnegative. This is indeed possible in a systematic manner, because the eigenvalues of $D^2 f(x) + \alpha I$ and $D^2 f(x)$ are very simply related (as we have already seen in the proof of

3.4 Convex Relaxation by the alphaBB Method

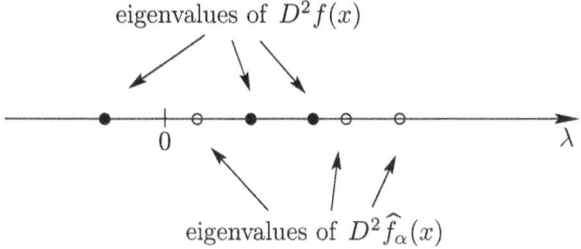

Fig. 3.11 Shifting of eigenvalues

Theorem 2.5.10b): Because of

$$\widehat{\lambda} \text{ eigenvalue of } D^2 \widehat{f_\alpha}(x) \Leftrightarrow \det(D^2 \widehat{f_\alpha}(x) - \widehat{\lambda} I) = 0$$
$$\Leftrightarrow \det(D^2 f(x) + \alpha I - \widehat{\lambda} I) = 0$$
$$\Leftrightarrow \det(D^2 f(x) - (\widehat{\lambda} - \alpha) I) = 0$$
$$\Leftrightarrow \widehat{\lambda} - \alpha \text{ eigenvalue of } D^2 f(x)$$

one obtains the eigenvalues of $D^2 \widehat{f_\alpha}(x)$ by shifting the eigenvalues of $D^2 f(x)$ to the right by α (Fig. 3.11). In particular, if $\lambda_{\min}$ denotes the *smallest* eigenvalue of $D^2 f(x)$, then $\lambda_{\min} + \alpha$ is the smallest eigenvalue of $D^2 \widehat{f_\alpha}(x)$, and we obtain

$$D^2 \widehat{f_\alpha}(x) \succeq 0 \Leftrightarrow \lambda_{\min} + \alpha \geq 0.$$

However, since $\lambda_{\min}$ also depends on x ($\lambda_{\min}(x)$ is the smallest eigenvalue of $D^2 f(x)$), one must state more precisely: $\widehat{f_\alpha}(x)$ is convex on X if and only if

$$\forall x \in X: \quad \lambda_{\min}(x) + \alpha \geq 0$$

holds, or equivalently

$$\min_{x \in X} \lambda_{\min}(x) + \alpha \geq 0$$

and thus

$$\alpha \geq -\min_{x \in X} \lambda_{\min}(x).$$

The attainment of this minimal value is due to the Weierstrass theorem and the fact that eigenvalues of symmetric matrices depend continuously on the matrix entries [39].

Overall, this leads to the following central result. It takes into account that for the relaxation property $\widehat{f_\alpha} \leq f$ we also assume $\alpha \geq 0$.

Theorem 3.4.3 *For all $\alpha \geq \max\{0, -\min_{x \in X} \lambda_{\min}(x)\}$ the function $\widehat{f}_\alpha = f + \alpha \psi$ is a convex relaxation of f on X.*

A fundamental problem for solving global optimization problems using convex relaxations by the αBB method is that for the appropriate choice of α one seemingly has to solve another global optimization problem, namely

$$\min \lambda_{\min}(x) \quad \text{s.t.} \quad x \in X.$$

This optimization problem is possibly even harder to solve than the original problem under consideration, so it seems that we are going in circles. At this point, interval arithmetic enters the stage and provides a guaranteed lower bound β for $\lambda_{\min}(x)$ on X, so that one avoids the solution of the global optimization problem. After the calculation of β, one sets

$$\alpha := \max\{0, -\beta\},$$

because this implies $\alpha \geq 0$ and $\alpha \geq -\beta \geq -\min_{x \in X} \lambda_{\min}(x)$. According to Theorem 3.4.3, $\widehat{f}_\alpha$ is therefore a convex relaxation of f on X.

The chosen α is generally larger than what can be achieved by global optimization of $\lambda_{\min}(x)$ over X, and therefore, according to Lemma 3.4.2c, one must accept a larger maximal distance between f and $\widehat{f}_\alpha$ on X. However, this choice of α is often easily implementable algorithmically.

We first calculate the lower bound β for two simple special cases before we turn to the general case.

Calculation of β for $n = 1$

In view of $D^2 f(x) = f''(x)$, the value $\lambda(x) = f''(x)$ is the only eigenvalue of $D^2 f(x)$, and this yields $\lambda_{\min}(x) = f''(x)$. If f'' is factorizable, for a given factorization we form the natural interval extension F'' of f'' and choose $\beta := \underline{F''}(X)$, thus defining β as the lower boundary point of $F''(X)$. Then $\beta \leq \min_{x \in X} f''(x) = \min_{x \in X} \lambda_{\min}(x)$ holds.

Example 3.4.4 The function $f(x) = x^3 - x$ is not convex on $X = [-1, 1]$ (Fig. 3.12). A convex relaxation of f on X is calculated by the αBB method as follows: Set $\widehat{f}_\alpha(x) = f(x) + \alpha \psi(x)$ with

$$\psi(x) = \frac{1}{2}(x+1)(x-1) = \frac{x^2 - 1}{2}$$

and

$$\alpha = \max\{0, -\beta\},$$

3.4 Convex Relaxation by the alphaBB Method

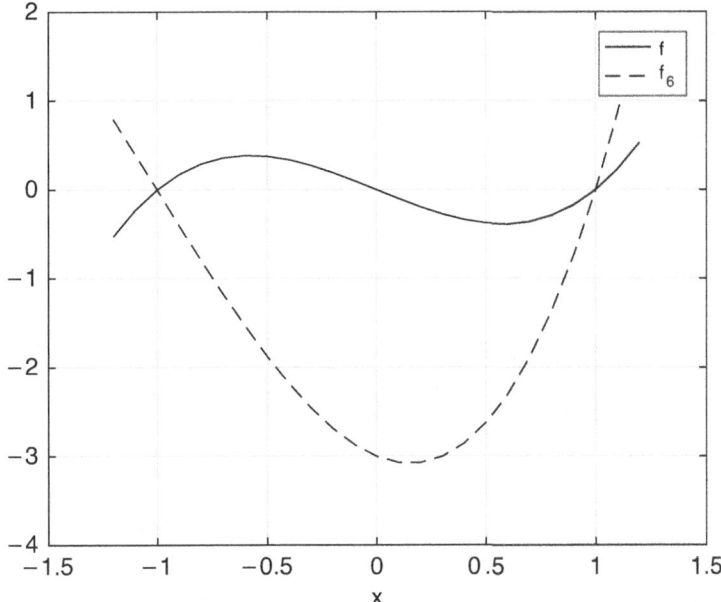

Fig. 3.12 Convex relaxation

where

$$\beta \leq \min_{x \in [-1,1]} f''(x) = \min_{x \in [-1,1]} 6x$$

is to be determined. We obtain $F''(X) = 6X$ and thus $F''([-1, 1]) = [-6, 6]$. It follows that $\beta = \underline{F}''([-1, 1]) = -6$ and $\alpha = 6$. Thus,

$$\widehat{f_6}(x) = f(x) + 6\psi(x) = x^3 - x + 3x^2 - 3$$

is a convex relaxation of f on $[-1, 1]$ (Fig. 3.12).

Exercise 3.4.5 Show that the function $\widehat{f_6}$ from Example 3.4.4 is neither convex nor a relaxation of f *outside* the interval $X = [-1, 1]$.

Calculation of β for $n > 1$ and Separable f

A function $f : \mathbb{R}^n \to \mathbb{R}$ is called *separable* if it can be written as the sum of n functions, each depending only on one of the n arguments:

$$f(x) = \sum_{i=1}^{n} f_i(x_i).$$

For example, $f_1(x) = x_1^2 + x_2^3 + e^{x_3}$ is separable, but $f_2(x) = x_1^2 x_2 + e^{x_3}$ is not. Separable functions satisfy

$$\nabla f(x) = \begin{pmatrix} f_1'(x_1) \\ \vdots \\ f_n'(x_n) \end{pmatrix}$$

and

$$D^2 f(x) = \begin{pmatrix} f_1''(x_1) & & 0 \\ & \ddots & \\ 0 & & f_n''(x_n) \end{pmatrix}$$

$$\text{(e.g. } D^2 f_1(x) = \begin{pmatrix} 2 & 0 & 0 \\ 0 & 6x_2 & 0 \\ 0 & 0 & e^{x_3} \end{pmatrix} \text{).}$$

For every separable C^2-function, the Hessian matrix $D^2 f(x)$ is therefore a diagonal matrix. Since in diagonal matrices the eigenvalues are identical to the diagonal elements, the smallest eigenvalue is

$$\lambda_{\min}(x) = \min_{i=1,\dots,n} f_i''(x_i).$$

Thus Exercises 1.3.3a and 1.3.6c imply

$$\min_{x \in X} \lambda_{\min}(x) = \min_{x \in X} \min_{i=1,\dots,n} f_i''(x_i) = \min_{i=1,\dots,n} \min_{x \in X} f_i''(x_i)$$

$$= \min_{i=1,\dots,n} \min_{x_i \in [\underline{x}_i, \overline{x}_i]} f_i''(x_i).$$

We therefore only need to calculate or estimate the smallest minimal value of n *one*-dimensional optimization problems. If all f_i'' are factorizable, we form natural interval extensions F_i'', choose $\beta_i := F_i''([\underline{x}_i, \overline{x}_i])$, $i = 1, \dots, n$, and set $\beta := \min_{i=1,\dots,n} \beta_i$. This implies $\beta \leq \min_{x \in X} \lambda_{\min}(x)$.

Calculation of β in the General Case

In the general case, the Hessian matrix of f is

$$D^2 f(x) = \begin{pmatrix} \partial_{x_1} \partial_{x_1} f(x) & \cdots & \partial_{x_n} \partial_{x_1} f(x) \\ \vdots & & \vdots \\ \partial_{x_1} \partial_{x_n} f(x) & \cdots & \partial_{x_n} \partial_{x_n} f(x) \end{pmatrix} =: A \quad \text{(actually: } A(x)\text{)}.$$

3.4 Convex Relaxation by the alphaBB Method

In this case, unfortunately, there are no explicit expressions for the eigenvalues anymore, but they are given only implicitly, as solutions of the equation $\det(A - \lambda I) = 0$. This is unfavorable for the application of interval arithmetic, because the latter provides bounds for the explicit evaluation of functions, not for implicit solutions of equations. Therefore, we will rather use explicit expressions for *bounds* on eigenvalues, to which secondary bounds are then calculated using interval arithmetic.

Definition 3.4.6 (Gershgorin Disks) For an (n,n)-matrix A with entries from the set of complex numbers $\mathbb{C}$, any $i \in \{1, \ldots, n\}$ and

$$r_i := \sum_{\substack{j=1 \\ j \neq i}}^{n} |a_{ij}|,$$

the set

$$\{\lambda \in \mathbb{C}| \, |\lambda - a_{ii}| \leq r_i\}$$

is called *Gershgorin disk* of A.

Theorem 3.4.7 (Gershgorin's Theorem) *Let A be an (n,n)-matrix with entries from $\mathbb{C}$. Then all eigenvalues of A lie in the set*

$$\bigcup_{i=1}^{n} \{\lambda \in \mathbb{C}| \, |\lambda - a_{ii}| \leq r_i\}.$$

Proof Let λ be some eigenvalue of A with corresponding eigenvector $v \in \mathbb{C} \setminus \{0\}$. Choose some $i \in \{1, \ldots, n\}$ with $|v_i| = \max_j |v_j|$. Then from the i-th row of the eigenvalue equation $Av = \lambda v$,

$$\sum_{j=1}^{n} a_{ij} v_j = \lambda v_i,$$

we obtain

$$\sum_{\substack{j=1 \\ j \neq i}}^{n} a_{ij} \frac{v_j}{|v_i|} = (\lambda - a_{ii}) \frac{v_i}{|v_i|}.$$

Taking absolute values on both sides and using the triangle inequality on the left side yields

$$\sum_{\substack{j=1 \\ j\neq i}}^{n} |a_{ij}| \frac{|v_j|}{|v_i|} \geq \left| \sum_{\substack{j=1 \\ j\neq i}}^{n} a_{ij} \frac{v_j}{|v_i|} \right| = \left| (\lambda - a_{ii}) \frac{v_i}{|v_i|} \right| = |\lambda - a_{ii}| \frac{|v_i|}{|v_i|}.$$

The maximality of $|v_i|$ therefore implies

$$\sum_{\substack{j=1 \\ j\neq i}}^{n} |a_{ij}| \geq |\lambda - a_{ii}|.$$

We have thus shown that for some i the inequality $|\lambda - a_{ii}| \leq r_i$ holds, which proves the assertion. $\square$

Example 3.4.8 For

$$A = \begin{pmatrix} 1 & 2 & 1 \\ 3 & 0 & 0 \\ -1 & 1 & -2 \end{pmatrix}$$

we have $r_1 = 3$, $r_2 = 3$ and $r_3 = 2$. The Gershgorin disks are therefore $\{\lambda \in \mathbb{C} | \, |\lambda - 1| \leq 3\}$, $\{\lambda \in \mathbb{C} | \, |\lambda| \leq 3\}$ and $\{\lambda \in \mathbb{C} | \, |\lambda + 2| \leq 2\}$ (Fig. 3.13). Indeed, the eigenvalues of A can be calculated to be $\lambda_1 = 3$ and $\lambda_{2/3} = -2 \pm i$, and they lie in the union of the three disks.

Since for $f \in C^2$ the Hessian matrix $A = D^2 f(x)$ not only has real entries, but is also symmetric, a result from linear algebra provides that all eigenvalues of A are *real* numbers [22]. The consideration of Gershgorin disks in the complex plane is

Fig. 3.13 Gershgorin disks

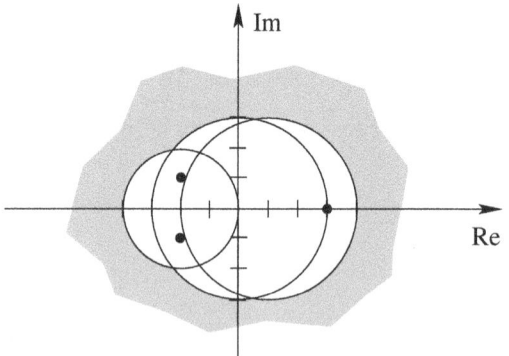

3.4 Convex Relaxation by the alphaBB Method

Fig. 3.14 Gershgorin intervals

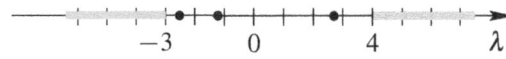

therefore not necessary for our purposes, but instead of the disks, we can use their intersections with the real axis, the *Gershgorin intervals* $[a_{ii} - r_i, a_{ii} + r_i]$.

Corollary 3.4.9 *Let A be a symmetric (n, n)-matrix with entries from $\mathbb{R}$. Then all eigenvalues of A lie in the set*

$$\bigcup_{i=1}^{n} [a_{ii} - r_i, a_{ii} + r_i].$$

Example 3.4.10 For the symmetric matrix

$$A = \begin{pmatrix} 1 & 2 & -1 \\ 2 & 0 & 0 \\ -1 & 0 & -2 \end{pmatrix}$$

we have $r_1 = 3$, $r_2 = 2$ and $r_3 = 1$. The Gershgorin intervals are therefore $[-2, 4]$, $[-2, 2]$ and $[-3, -1]$ (Fig. 3.14). The actual eigenvalues are $\lambda_1 \approx -2.5$, $\lambda_2 \approx -1.2$ and $\lambda_3 \approx 2.7$.

Exercise 3.4.11 Would it be correct in Corollary 3.4.9 to replace the set $\bigcup_{i=1}^{n} [a_{ii} - r_i, a_{ii} + r_i]$ by the interval $[\min_{i=1,\dots,n} (a_{ii} - r_i), \max_{i=1,\dots,n} (a_{ii} + r_i)]$? If so, why might this not be advisable?

In any case, a lower bound for all eigenvalues, and thus also for $\lambda_{\min}$, is the smallest of the interval boundary points:

$$\lambda_{\min} \geq \min_{i=1,\dots,n} (a_{ii} - r_i).$$

If we again consider the x-dependence, i.e. $A(x) = D^2 f(x)$, it follows

$$\forall x \in X: \quad \lambda_{\min}(x) \geq \min_{i=1,\dots,n} (a_{ii}(x) - r_i(x))$$

with

$$r_i(x) = \sum_{\substack{j=1 \\ j \neq i}}^{n} |a_{ij}(x)|, \quad i = 1, \dots, n.$$

In view of Exercise 1.3.3a this yields

$$\min_{x \in X} \lambda_{\min}(x) \geq \min_{x \in X} \min_{i=1,\ldots,n} (a_{ii}(x) - r_i(x)) = \min_{i=1,\ldots,n} \min_{x \in X} (a_{ii}(x) - r_i(x)).$$

This is the desired functional expression for a lower bound of the smallest eigenvalue of $D^2 f$ on X.

Using interval arithmetic, next we generate a further, but computationally manageable lower bound as follows. If all entries a_{ij} of $D^2 f$ are factorizable, we form natural interval extensions A_{ij} as well as interval extensions R_i of r_i, choose for each $i = 1, \ldots, n$

$$\beta_i := \underline{A_{ii}(X) - R_i(X)}$$

and set $\beta := \min_{i=1,\ldots,n} \beta_i$. Then $\beta \leq \min_{x \in X} \lambda_{\min}(x)$ holds.

As a natural interval extension of the absolute value function $\mathrm{abs}(x) := |x|$ on $\mathbb{R}$ we use

$$\mathrm{ABS}([\underline{x}, \overline{x}]) = \begin{cases} [\min\{|\underline{x}|, |\overline{x}|\}, \max\{|\underline{x}|, |\overline{x}|\}], & \text{if } 0 \notin [\underline{x}, \overline{x}] \\ [0, \max\{|\underline{x}|, |\overline{x}|\}], & \text{if } 0 \in [\underline{x}, \overline{x}], \end{cases}$$

so

$$A_{ii}(X) - R_i(X) = A_{ii}(X) - \sum_{\substack{j=1 \\ j \neq i}}^{n} \mathrm{ABS}(A_{ij}(X)).$$

Thus, in explicit terms we obtain for each $i = 1, \ldots, n$

$$\beta_i = \underline{A_{ii}(X) - R_i(X)} = \underline{A_{ii}(X)} - \overline{R_i(X)}$$

$$= \underline{A_{ii}(X)} - \sum_{\substack{j=1 \\ j \neq i}}^{n} \overline{\mathrm{ABS}(A_{ij}(X))}$$

$$= \underline{A_{ii}(X)} - \sum_{\substack{j=1 \\ j \neq i}}^{n} \max\left\{|\underline{A_{ij}(X)}|, |\overline{A_{ij}(X)}|\right\},$$

and we have shown the following theorem.

3.4 Convex Relaxation by the alphaBB Method

Theorem 3.4.12 (αBB Relaxation of a Function) *For $f \in C^2(X, \mathbb{R})$ let all entries a_{ij} of $D^2 f$ be factorizable with natural interval extensions A_{ij}. Then with*

$$\beta_i := \underline{A}_{ii}(X) - \sum_{\substack{j=1 \\ j \neq i}}^{n} \max\left\{|\underline{A}_{ij}(X)|, |\overline{A}_{ij}(X)|\right\}, \quad i = 1, \ldots, n,$$

$$\beta := \min_{i=1,\ldots,n} \beta_i,$$

$$\alpha := \max\{0, -\beta\}$$

the function $\widehat{f}_\alpha := f + \alpha \psi$ is a convex relaxation of f on X.

The calculation from Theorem 3.4.12 may indeed yield the value $\alpha = 0$. Then it has been algorithmically proven that f coincides with its convex relaxation $\widehat{f}_0$, thus being a convex function on X itself. This can be useful if the convexity of f is not expected, but also not excluded. After such an algorithmic proof of convexity, the minimization of f on X can be performed using the techniques from Chap. 2.

Example 3.4.13 The function $f(x) = x_1 \exp(x_2)$ is twice continuously differentiable on $X = [-1, 1] \times [0, 1]$ and possesses the Hessian

$$D^2 f(x) = \begin{pmatrix} 0 & \exp(x_2) \\ \exp(x_2) & x_1 \exp(x_2) \end{pmatrix}$$

with factorizable entries. Their natural interval extensions are

$$A_{11}(X) = 0,$$
$$A_{12}(X) = A_{21}(X) = \text{EXP}(X_2),$$
$$A_{22}(X) = X_1 \text{EXP}(X_2).$$

We thus obtain

$$\begin{aligned}
\beta_1 &= \underline{A}_{11}(X) - \max\left\{|\underline{A}_{12}(X)|, |\overline{A}_{12}(X)|\right\} \\
&= 0 - \max\left\{|\underline{\text{EXP}}([0, 1])|, |\overline{\text{EXP}}([0, 1])|\right\} \\
&= -\max\{|\exp(0)|, |\exp(1)|\} = -e, \\
\beta_2 &= \underline{A}_{22}(X) - \max\left\{|\underline{A}_{21}(X)|, |\overline{A}_{21}(X)|\right\} \\
&= \underline{[-1, 1] \text{EXP}([0, 1])} - \max\left\{|\underline{\text{EXP}}([0, 1])|, |\overline{\text{EXP}}([0, 1])|\right\}
\end{aligned}$$

$$= [-1, 1][1, e] - e = -2e,$$
$$\beta = \min\{\beta_1, \beta_2\} = -2e,$$
$$\alpha = \max\{0, -\beta\} = 2e$$

and, by Theorem 3.4.12, the convex relaxation

$$\widehat{f}_{2e}(x) = x_1 \exp(x_2) + e\,(x_1^2 + x_2^2 - x_2 - 1)$$

of f on X.

3.5 Uniformly Refined Tessellations

Next we turn to the question of how to use convex relaxations of twice continuously differentiable functions on boxes, as constructed in Sect. 3.4, for the algorithmic solution of global optimization problems.

Example 3.5.1 Once again we consider the function $f(x) = x^3 - x$ on $X = [-1, 1]$ from Example 3.4.4 and denote the global minimal value of

$$P: \quad \min f(x) \quad \text{s.t.} \quad x \in X$$

with v. Instead of calculating v using optimality conditions, we will try to provide good *bounds* on v in the following.

Upper Bound
Upper bounds on optimal values may by obtained by evaluating the objective function at any feasible point, as it holds

$$\forall x \in X: \quad v \leq f(x).$$

In the present example one possibility for this, although possibly not the most clever one, is $v \leq f(0) = 0$.

Lower Bound
According to Theorem 3.2.5a $v \geq \widehat{v}$ holds for the minimal value $\widehat{v}$ of

$$\widehat{P}: \quad \min \widehat{f}(x) \quad \text{s.t.} \quad x \in X$$

with any convex relaxation $\widehat{f}$ of f on X (since the feasible set X is already a convex set). In Example 3.4.4, we saw that $\widehat{f}(x) = x^3 + 3x^2 - x - 3$ is such a convex relaxation of f on X. The critical points of $\widehat{f}$ on all of $\mathbb{R}$ are calculated from

$$0 = \widehat{f}'(x) = 3x^2 + 6x - 1$$

3.5 Uniformly Refined Tessellations

to

$$x_{1/2} = -1 \pm \sqrt{1 + \frac{1}{3}} = -1 \pm \frac{2}{\sqrt{3}}.$$

The only critical point in X is thus $\widehat{x} = 2/\sqrt{3} - 1$ with $\widehat{f}(\widehat{x}) > -3.08$. Thus, we have found a KKT point of the convex problem $\widehat{P}$ and no longer need to examine the boundary points of X. The point $\widehat{x} = 2/\sqrt{3} - 1$ is thus the global minimal point of $\widehat{P}$ with $\widehat{v} > -3.08$. This provides the lower bound $v \geq -3.08$ for v.

Overall, we thus obtain the enclosure $v \in [-3.08, 0]$ for the optimal value of P.

Instead of determining an upper bound on v using some *arbitrary* point $x \in X$ (like $x = 0$ in Example 3.5.1), there are better alternatives:

- By a method of nonlinear optimization [37] determine a local minimal point x^{loc} of f on X. Then $v \leq v^{\text{loc}} := f(x^{\text{loc}})$ holds, and at least one can hope to achieve a good upper bound for v. In Example 3.5.1 this approach yields the following improvement: From $0 = f'(x) = 3x^2 - 1$ follows after the usual calculations for example $x^{\text{loc}} = 1/\sqrt{3}$ with $v^{\text{loc}} = (1/\sqrt{3})^3 - 1/\sqrt{3} < -0.38$, thus the enclosure $v \in [-3.08, -0.38]$. However, there is also the further local minimal point $x^{\text{loc}} = -1$, which leads to the worse upper bound $v^{\text{loc}} = (-1)^3 - (-1) = 0$. Unfortunately, it cannot be guaranteed that one finds a 'good' local minimal point by a method of nonlinear optimization.
- Insert $\widehat{x}$ into f, hoping that a global minimal point of the relaxation also provides a low value for the original function:

$$v \leq f(\widehat{x}).$$

In Example 3.5.1 this leads to $(2/\sqrt{3} - 1)^3 - (2/\sqrt{3} - 1) < -0.15$ and thus to the enclosure $v \in [-3.08, -0.15]$.

While with the first of the above alternatives one often finds a good upper bound for v with some effort (application of a method of nonlinear optimization), the advantages of the second alternative are that it is easy to implement (namely by a single function evaluation) and that αBB relaxations are accompanied by information about the distance between upper and lower bound. Indeed,

$$v \in [\widehat{v}, f(\widehat{x})] = [\widehat{f}_\alpha(\widehat{x}), f(\widehat{x})]$$

holds with

$$w([\widehat{v}, f(\widehat{x})]) = f(\widehat{x}) - \widehat{f}_\alpha(\widehat{x}) \leq \max_{x \in X}(f(x) - \widehat{f}_\alpha(x)) \stackrel{\text{Lemma 3.4.2c}}{=} \frac{\alpha}{8} w(X)^2.$$

This shows the following result.

Theorem 3.5.2 *Let $X \in \mathbb{R}^n$, $f \in C^2(X, \mathbb{R})$, $D^2 f$ be factorizable, v be the global minimal value of*

$$P: \quad \min f(x) \quad \text{s.t.} \quad x \in X,$$

$\widehat{f}_\alpha$ *be a convex relaxation of f on X, constructed by Theorem 3.4.12, $\widehat{x}$ be a global minimal point of $\widehat{f}_\alpha$ on X, and $\widehat{v}$ be the minimal value $\widehat{f}_\alpha(\widehat{x})$. Then the enclosure*

$$v \in [\widehat{v}, f(\widehat{x})] \quad \text{with} \quad w([\widehat{v}, f(\widehat{x})]) \leq \frac{\alpha}{8} w(X)^2$$

holds.

Theorem 3.5.2 implies better bounds for v, the smaller the box X is, and this is *independent* of how coarsely α is chosen. For $w(X) \to 0$, even $\widehat{v} \nearrow v$ applies (due to the convexity of X and Theorem 3.2.5d, this does *not* contradict Theorem 3.2.5c).

Since X is fixed, we may not change the size of X, but we can *subdivide* X into smaller boxes (Fig. 3.15). In the following definition, int$A$ denotes the topological interior of a set A.

Fig. 3.15 Tessellation of a box

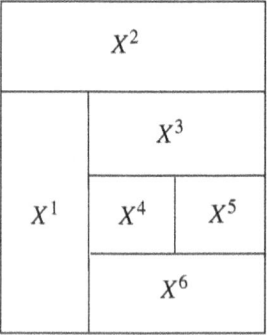

3.5 Uniformly Refined Tessellations

Definition 3.5.3 (Tessellation) For $k \in \mathbb{N}$, the boxes $X^1, \ldots, X^k \in \mathbb{R}^n$ form a *tessellation* of the box $X \in \mathbb{R}^n$, if

(a) $\bigcup_{\ell=1}^{k} X^\ell = X$ and
(b) $\forall j \neq \ell : (\text{int} X^j) \cap (\text{int} X^\ell) = \emptyset$

apply. For $n = 2$ a tessellation is also called *tiling*.

For every tessellation of X, there is (at least) one sub-box $X^{\ell_{\text{glob}}}$ that contains a global minimal point x^{glob} of f on X. The global minimal value is found by comparing all minimal values of f on X^ℓ, $\ell = 1, \ldots, k$:

$$v = \min_{x \in X} f(x) = \min_{x \in \bigcup_{\ell=1}^{k} X^\ell} f(x) = \min_{\ell=1,\ldots,k} \min_{x \in X^\ell} f(x),$$

where we used Exercise 1.3.4. With the definition $v^\ell := \min_{x \in X^\ell} f(x)$, we can also write this as $v = \min_{\ell=1,\ldots,k} v^\ell$, which is illustrated in Fig. 3.16 for $n = 1$ and $k = 4$.

The basic idea of the following is to calculate lower bounds $\widehat{v}^\ell$ for v^ℓ using the αBB method on *each* sub-box X^ℓ. From $v = \min_{\ell=1,\ldots,k} v^\ell$ and $v^\ell \geq \widehat{v}^\ell$, $\ell = 1, \ldots, k$, it then follows

$$v \geq \min_{\ell=1,\ldots,k} \widehat{v}^\ell.$$

Since the sub-boxes X^ℓ, $\ell = 1, \ldots, k$, are smaller than X, this lower bound on v should be better (i.e., larger) than some $\widehat{v}$ calculated as in Theorem 3.5.2 for the entire box X.

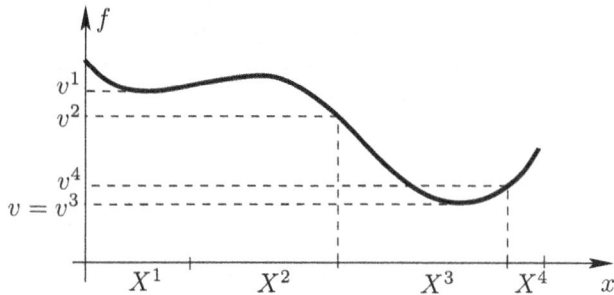

Fig. 3.16 Minimal values on sub-boxes

For the calculation of the values $\widehat{v}^\ell$, let

$$\forall \ell = 1, \ldots, k: \quad \widehat{f}^\ell_{\alpha_\ell}(x) := f(x) + \frac{\alpha_\ell}{2}(\underline{x}^\ell - x)^\mathsf{T}(\overline{x}^\ell - x)$$

with

$$\alpha_\ell \geq \max\left\{0, -\min_{x \in X^\ell} \lambda_{\min}(x)\right\}$$

be a convex relaxation of f on X^ℓ.

If some $\alpha \geq \max\{0, -\min_{x \in X} \lambda_{\min}(x)\}$ is known, then one may also set $\alpha_\ell := \alpha$ for all $\ell = 1, \ldots, k$, because one obtains

$$X^\ell \subseteq X \Rightarrow \min_{x \in X^\ell} \lambda_{\min}(x) \geq \min_{x \in X} \lambda_{\min}(x)$$

$$\Rightarrow \max\left\{0, -\min_{x \in X^\ell} \lambda_{\min}(x)\right\} \leq \max\left\{0, -\min_{x \in X} \lambda_{\min}(x)\right\} \leq \alpha.$$

In the following, for simplicity, we use this choice ($\alpha_\ell := \alpha$ for all ℓ), although the determination of new α_ℓ for all X^ℓ could lead to better bounds. So let

$$\widehat{f}^\ell_\alpha(x) = f(x) + \frac{\alpha}{2}(\underline{x}^\ell - x)^\mathsf{T}(\overline{x}^\ell - x),$$

$\widehat{x}^\ell$ be a global minimal point of $\widehat{f}^\ell_\alpha$ on X^ℓ, and $\widehat{v}^\ell$ the associated minimal value. Then, according to Theorem 3.5.2, for all $\ell = 1, \ldots, k$ we have

$$v^\ell \in [\widehat{v}^\ell, f(\widehat{x}^\ell)] \quad \text{with} \quad w([\widehat{v}^\ell, f(\widehat{x}^\ell)]) \leq \frac{\alpha}{8} w(X^\ell)^2.$$

As already seen above, the expression $\min_{\ell=1,\ldots,k} \widehat{v}^\ell$ forms a lower bound for v. To examine it more closely, we focus on some index ℓ_* at which the smallest of the values $\widehat{v}^\ell$ is attained. So let $\widehat{v}^{\ell_*} = \min_{\ell=1,\ldots,k} \widehat{v}^\ell$ and $\widehat{x}^{\ell_*}$ be a global minimal point of $\widehat{f}^{\ell_*}_\alpha$ on X^{ℓ_*}. Then $v \geq \widehat{v}^{\ell_*}$ and, due to $\widehat{x}^{\ell_*} \in X$, also $v \leq f(\widehat{x}^{\ell_*})$ hold (Fig. 3.17 illustrates that $v = v^{\ell_*}$ is not necessarily true, i.e., X^{ℓ_*} does not necessarily need to contain a global minimal point of f on X). In total, we obtain

$$v \in [\widehat{v}^{\ell_*}, f(\widehat{x}^{\ell_*})] \quad \text{with} \quad w([\widehat{v}^{\ell_*}, f(\widehat{x}^{\ell_*})]) \leq \frac{\alpha}{8} w(X^{\ell_*})^2.$$

Therefore, we are able to *control* the accuracy of the calculation of v (i.e., the length of the enclosure interval for v) by the box sizes: With a user-specified tolerance $\varepsilon > 0$ the accuracy $w([\widehat{v}^{\ell_*}, f(\widehat{x}^{\ell_*})]) \leq \varepsilon$ is guaranteed when, in the case $\alpha > 0$, the box width of X^{ℓ_*} can be bounded in the form

$$w(X^{\ell_*}) \leq \sqrt{\frac{8\varepsilon}{\alpha}}.$$

3.5 Uniformly Refined Tessellations

Fig. 3.17 Relaxations on sub-boxes

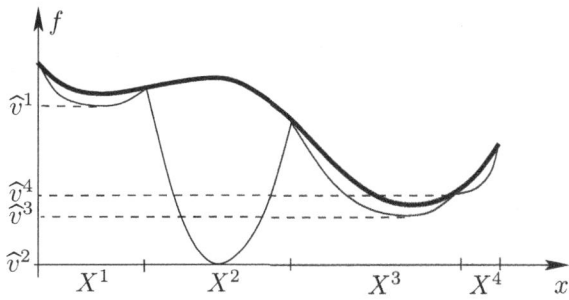

In the (uninteresting) case $\alpha = 0$, the box width does not matter, and every tolerance $\varepsilon > 0$ is maintained.

A small length of the enclosure interval for v implies the following further result.

Lemma 3.5.4 *Let the value $\widehat{v}^{\ell_*}$ and the point $\widehat{x}^{\ell_*}$ be calculated as above, and for some $\varepsilon > 0$ let $w([\widehat{v}^{\ell_*}, f(\widehat{x}^{\ell_*})]) \leq \varepsilon$. Then the point $\widetilde{x} := \widehat{x}^{\ell_*}$ satisfies the conditions*

$$\widetilde{x} \in X \quad \text{and} \quad v \leq f(\widetilde{x}) \leq v + \varepsilon$$

and is hence an ε-optimal feasible point of P.

Proof In view of $\widetilde{x} = \widehat{x}^{\ell_*} \in X^{\ell_*} \subseteq X$ the conditions $\widetilde{x} \in X$ and $v \leq f(\widetilde{x})$ hold. From

$$\varepsilon \geq w([\widehat{v}^{\ell_*}, f(\widehat{x}^{\ell_*})]) = f(\widetilde{x}) - \widehat{v}^{\ell_*} \geq f(\widetilde{x}) - v$$

also $f(\widetilde{x}) \leq v + \varepsilon$ follows. □

If we could implement the preceding ideas algorithmically, we would be able to generate an enclosure interval of maximal length ε for v and an ε-optimal feasible point in the sense of Lemma 3.5.4 for any given tolerance ε. Nothing more may be expected from a numerical algorithm for global optimization.

The central remaining problem is the generation of a sufficiently small box X^{ℓ_*}. For its construction we first pursue the approach of tessellating X *uniformly* into sub-boxes $X^1, \ldots, X^k$, for example by recursively subdividing the boxes at their centers $m(X^\ell) = (\underline{x}^\ell + \overline{x}^\ell)/2$, $\ell = 1, \ldots, k$.

Fig. 3.18 Uniform tessellation at box midpoints.

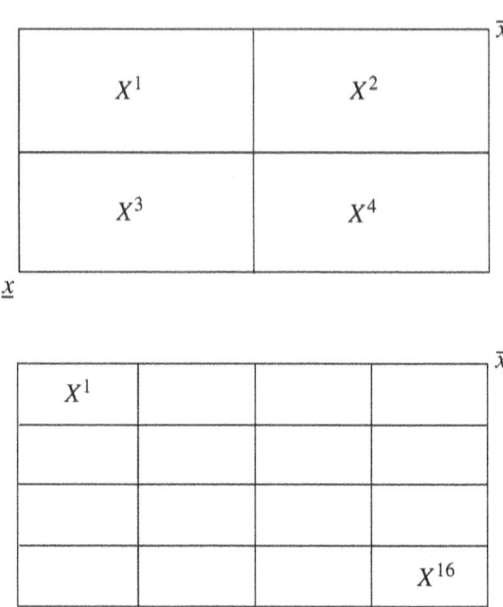

Example 3.5.5 For $n = 2$, we consider the box $X = [\underline{x}, \overline{x}]$ in Fig. 3.18. After a uniform subdivision step at the box center, the four resulting sub-boxes satisfy

$$\forall \ell = 1, \ldots, 4: \quad w(X^\ell) = \frac{w(X)}{2}$$

and after a further refinement step

$$\forall \ell = 1, \ldots, 16: \quad w(X^\ell) = \frac{w(X)}{4}.$$

After r refinement steps, all boxes possess the width $w(X)/2^r$. In particular, the box X^{ℓ_*} satisfies

$$f(\widehat{x}^{\ell_*}) - \widehat{v}^{\ell_*} \leq \frac{\alpha}{8} w(X^{\ell_*})^2 = \frac{\alpha}{8} \frac{w(X)^2}{4^r}.$$

To achieve a tolerance $\varepsilon > 0$ in the calculation of v, it is thus sufficient (in the case $\alpha > 0$) to refine sufficiently often, i.e., to choose r sufficiently large:

$$\frac{\alpha}{8} \frac{w(X)^2}{4^r} \stackrel{!}{\leq} \varepsilon \quad \Leftrightarrow \quad \left(\frac{1}{4}\right)^r \leq \frac{8\varepsilon}{\alpha w(X)^2}.$$

3.5 Uniformly Refined Tessellations

$$\Leftrightarrow \quad r \geq \log\left(\frac{8\varepsilon}{\alpha w(X)^2}\right) / \log\left(\frac{1}{4}\right). \tag{3.2}$$

Note that this lower bound for r may be nonpositive, namely for $\varepsilon \geq \alpha w(X)^2/8$. In this case the termination tolerance ε is so coarse that, in spite of $\alpha > 0$, no subdivision is necessary to meet the required accuracy.

Keeping this effect in mind, the minimal $r \in \mathbb{N}_0$ with (3.2) is

$$r = \left\lceil \log\left(\frac{8\varepsilon}{\alpha w(X)^2}\right) / \log\left(\frac{1}{4}\right) \right\rceil^+,$$

where $\lceil a \rceil$ denotes the *upper Gaussian bracket* of $a \in \mathbb{R}$, i.e., the smallest $z \in \mathbb{Z}$ with $z \geq a$, and $a^+ := \max\{0, a\}$ is the *positive part* of a.

It is easy to see that the calculation of r in Example 3.5.5 does not depend on the dimension n. Thus, we can formulate Algorithm 3.2 for the global minimization of f on X.

Algorithm 3.2: Global minimization of a box-constrained function by uniform tessellation refinement

---

Input: $X = [\underline{x}, \overline{x}] \in \mathbb{R}^n$, $f \in C^2(X, \mathbb{R})$ with factorizable Hessian matrix $D^2 f$, termination tolerance $\varepsilon > 0$
Output: ε-optimal feasible point $\widetilde{x}$ of f on X, i.e. $\widetilde{x} \in X$ with $v \leq f(\widetilde{x}) \leq v + \varepsilon$

1 begin
2   Compute some $\alpha \geq \max\{0, -\min_{x \in X} \lambda_{\min}(x)\}$.
3   Set

$$r = \begin{cases} \left\lceil \log\left(\frac{8\varepsilon}{\alpha w(X)^2}\right) / \log\left(\frac{1}{4}\right) \right\rceil^+, & \text{if } \alpha > 0 \\ 0, & \text{if } \alpha = 0 \end{cases}$$

   and subdivide X r times uniformly into sub-boxes $X^1, \ldots, X^k$.
4   For each $\ell = 1, \ldots, k$ compute the minimal value $\widehat{v}^\ell$ of

$$\widehat{f}_\alpha^\ell(x) = f(x) + \frac{\alpha}{2}(x^\ell - x)^\mathsf{T}(\overline{x}^\ell - x)$$

   on X^ℓ (e.g., with a method from Sect. 2.8).
5   Choose some ℓ_* with $\widehat{v}^{\ell_*} = \min_{\ell=1,\ldots,k} \widehat{v}^\ell$.
6   If not already done in line 4, compute a minimal point $\widehat{x}^{\ell_*}$ of $\widehat{f}_\alpha^{\ell_*}$ on X^{ℓ_*}.
7   Set $\widetilde{x} := \widehat{x}^{\ell_*}$.
8 end

Algorithm 3.2 is rather simple, but in general not advisable in practice, as the number k of boxes in line 3 increases exponentially: We have $k = (2^n)^r$. For small tolerances ε, one would therefore have to solve a usually unmanageable large number of subproblems in line 4.

Example 3.5.6 For $n = 3$ we consider the box

$$X = \left[-\begin{pmatrix} 1 \\ 1 \\ 1 \end{pmatrix}, \begin{pmatrix} 1 \\ 1 \\ 1 \end{pmatrix} \right]$$

with width $w(X) = 2\sqrt{3}$. For the (not uncommon) parameter choices $\alpha = 1$ and $\varepsilon = 10^{-3}$, we then obtain

$$r = \left\lceil \log\left(\frac{8\varepsilon}{\alpha w(X)^2}\right) / \log\left(\frac{1}{4}\right) \right\rceil^+ = 6.$$

Thus, the execution of Algorithm 3.2 requires a tessellation of X into $(2^3)^6 = 262.144$ sub-boxes.

3.6 Branch-and-Bound for Box-Constrained Problems

In this section, we provide a *practical* method that generates an ε-optimal feasible point of P in usually 'manageably many' steps for box-constrained problems

$$P: \quad \min f(x) \quad \text{s.t.} \quad x \in X$$

with $X \in \mathbb{R}^n$, $f \in C^2(X, \mathbb{R})$, factorizable second derivative $D^2 f$ and for any tolerance $\varepsilon > 0$. It is based on the considerations from Sect. 3.5, but without constructing the tessellation of X by uniform refinement.

The main motivation for the following is to keep the number of sub-boxes as small as possible. Essential strategies for this are:

- In each step, only *one* sub-box is refined, and it is only *halved* (branching).
- The calculation of bounds on the sub-boxes (bounding) allows the 'most promising' choice of the sub-box to be divided as well as the exclusion of boxes which cannot contain better points than the already known ones (discarding, pruning, or fathoming). In contrast to the *uniform* refinement from Sect. 3.5, this leads to an *adaptive* refinement of the tessellation.

Next we will consider these strategies in more detail.

Halving of Boxes
Let the box $X^\ell = [\underline{x}^\ell, \overline{x}^\ell]$ with $\ell \in \{1, \ldots, k\}$ to be divided into two equally sized sub-boxes. To this end, we define the one-dimensional intervals $X_i^\ell = [\underline{x}_i^\ell, \overline{x}_i^\ell]$,

3.6 Branch-and-Bound for Box-Constrained Problems

Fig. 3.19 Halvings of a box

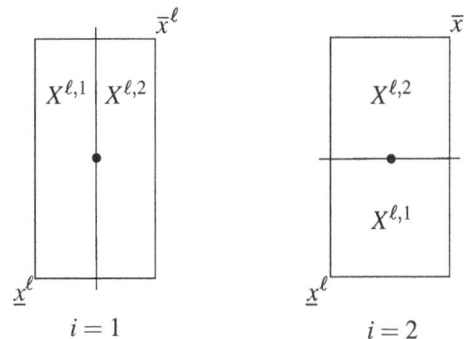

$i = 1, \ldots, n$, so that

$$X^\ell = X_1^\ell \times \ldots \times X_n^\ell \tag{3.3}$$

holds. Let us choose some $i \in \{1, \ldots, n\}$ and define $X^{\ell,1}$ as the box that results from (3.2) when X_i^ℓ is replaced by the interval $[\underline{x}_i^\ell, m(X_i^\ell)]$, as well as $X^{\ell,2}$ as the box with X_i^ℓ replaced by $[m(X_i^\ell), \overline{x}_i^\ell]$. Written out this means

$$X^{\ell,1} = \left[\begin{pmatrix} \underline{x}_1^\ell \\ \vdots \\ \underline{x}_i^\ell \\ \vdots \\ \underline{x}_n^\ell \end{pmatrix}, \begin{pmatrix} \overline{x}_1^\ell \\ \vdots \\ \frac{\underline{x}_i^\ell + \overline{x}_i^\ell}{2} \\ \vdots \\ \overline{x}_n^\ell \end{pmatrix} \right], \quad X^{\ell,2} = \left[\begin{pmatrix} \underline{x}_1^\ell \\ \vdots \\ \frac{\underline{x}_i^\ell + \overline{x}_i^\ell}{2} \\ \vdots \\ \underline{x}_n^\ell \end{pmatrix}, \begin{pmatrix} \overline{x}_1^\ell \\ \vdots \\ \overline{x}_i^\ell \\ \vdots \\ \overline{x}_n^\ell \end{pmatrix} \right].$$

Figure 3.19 shows the halving of a box $X^\ell \in \mathbb{R}^2$ with the choices $i = 1$ and $i = 2$.

Depending on the selection rule for $i \in \{1, \ldots, n\}$, the new boxes can be 'lengthy' (when halving along the shortest edge) or 'rather cube-like' (when halving along the longest edge). However, by successive halving along, e.g., the shortest edge, we cannot ensure that the *widths* of the resulting boxes tend to zero. In the following, we will therefore halve boxes along the (or one) longest edge, thus choosing some $i \in \{1, \ldots, n\}$ with

$$\overline{x}_i^\ell - \underline{x}_i^\ell = \max_{j=1,\ldots,n} \left(\overline{x}_j^\ell - \underline{x}_j^\ell \right) \quad (= \|\overline{x}^\ell - \underline{x}^\ell\|_\infty).$$

We will see in Lemma 3.6.1 to which reduction of the box widths this halving rule leads.

How boxes are successively divided can be illustrated by a branching tree. Figure 3.20 shows the division of the box X into the two sub-boxes X^1 and X^2, of which X^1 is again divided into $X^{1,1}$ and $X^{1,2}$ and $X^{1,2}$ into $X^{1,2,1}$ and $X^{1,2,2}$.

Fig. 3.20 Branching tree

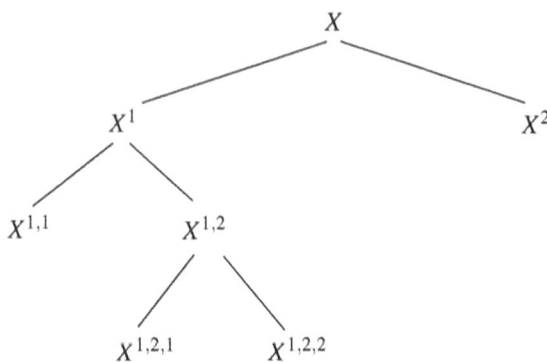

At a later point in the procedure, X^2 could also be divided, etc. We will not need a precise addressing of the nodes of the tree as in Fig. 3.20, but only the fact that the branching tree is binary, i.e., each node has at most two children and exactly one parent node (except for the root node, which corresponds to the box X). The deeper one descends down the tree, the smaller the sub-boxes become.

Bounds

The main idea for reducing the number of considered sub-boxes is based on the knowledge of some feasible reference point $\tilde{x} \in X$, such as the box center $m(X)$. Its objective function value $\tilde{v} = f(\tilde{x})$ forms an upper bound for v. More importantly, the value $\tilde{v}$ can also be used to exclude certain sub-boxes X^ℓ from further consideration. Indeed, the relation $\tilde{v} \leq v^\ell$ implies

$$f(\tilde{x}) = \tilde{v} \leq v^\ell = \min_{x \in X^\ell} f(x)$$

and thus $f(\tilde{x}) \leq f(x)$ for all $x \in X^\ell$. Therefore, the sub-box X^ℓ does not contain any point x with a better objective function value than $\tilde{x}$ and, therefore, does not need to be considered further. In the worst case, $\tilde{x}$ is already a (not yet recognized as such) global minimal point, and we discard a sub-box X^ℓ that may contain another global minimal point. However, since we do not want to approximate the set of all global minimal points, but only one of them, we may proceed as described.

Instead of the inequality $\tilde{v} \leq v^\ell$ involving the algorithmically inaccessible value v^ℓ, also the relation $\tilde{v} \leq \hat{v}^\ell$ can be used as a sufficient criterion for discarding the box X^ℓ, because $\hat{v}^\ell \leq v^\ell$ implies $\tilde{v} \leq v^\ell$ and thus the above arguments.

For the function f in Fig. 3.21, the box X^{ℓ_1} can therefore be discarded. Neither the box X^{ℓ_2} contains any points x with a better objective function value than $\tilde{x}$, but this cannot be determined based on the lower bound $\hat{v}^{\ell_2}$, since it is too coarse for this purpose.

The smaller the bound $\tilde{v}$, the more sub-boxes may be discarded. Therefore, it is adjusted each time a feasible point $x' \in X$ with a better objective function value than $\tilde{x}$, i.e., $f(x') < f(\tilde{x}) = \tilde{v}$, is generated. The previously best known feasible point

3.6 Branch-and-Bound for Box-Constrained Problems

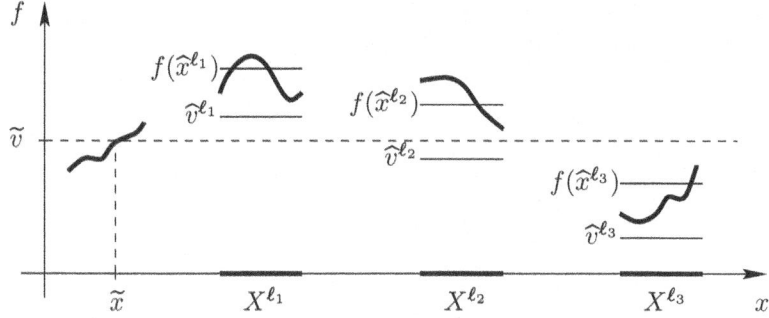

Fig. 3.21 Bounds in branch-and-bound methods

$\widetilde{x}$ is then replaced by the new reference point x' and $\widetilde{v}$ by $f(x')$. In the situation of Fig. 3.21, one may set $x' := \widehat{x}^{\ell_3}$. After the corresponding adjustment of $\widetilde{v}$ to $f(\widehat{x}^{\ell_3})$, it becomes possible to also discard the box X^{ℓ_2}. In view of this update procedure, the reference points $\widetilde{x}$ are also called *incumbents*.

Another effect of this adjustment of the value of $\widetilde{v}$ is that it forms an increasingly accurate upper bound for v. To also obtain a lower bound for v, we calculate the smallest of all minimal values $\widehat{v}^{\ell}$ of $\widehat{f}_{\alpha}^{\ell}$ on X^{ℓ} for the given tessellation $X^1, \ldots, X^k$, i.e., we determine some $\ell_\star$ with $\widehat{v}^{\ell_\star} = \min_{\ell=1,\ldots,k} \widehat{v}^{\ell}$. This yields $v \in [\widehat{v}^{\ell_\star}, \widetilde{v}]$. If the length $w([\widehat{v}^{\ell_\star}, \widetilde{v}]) = \widetilde{v} - \widehat{v}^{\ell_\star}$ of this enclosing interval for v drops below a given tolerance $\varepsilon > 0$, the procedure terminates. Otherwise, we try to improve the bounds.

To improve the lower bound $\widehat{v}^{\ell_\star}$ on v, it suggests itself to halve the box $X^{\ell_\star}$, which is responsible for the 'bad' minimal value $\widehat{v}^{\ell_\star}$. As a welcome side effect, the solution of the relaxations of the two new subproblems also provide new feasible points, which in turn can improve the upper bound $\widetilde{v}$. We will see in Theorem 3.6.2 that the length of the enclosure interval $[\widehat{v}^{\ell_\star}, \widetilde{v}]$ actually falls below any given tolerance $\varepsilon > 0$ after a finite number of subdivision steps.

Algorithm 3.3 implements these considerations by maintaining a list of boxes from the current tessellation in which a better than the best known feasible point $\widetilde{x}$ may still lie. In order to be able to perform the necessary comparisons, the sub-boxes X^{ℓ} are stored together with the associated bound $\widehat{v}^{\ell}$ as a pair $(X^{\ell}, \widehat{v}^{\ell})$ in the list.

The fact that the list in Algorithm 3.3 is completely emptied, as queried in lines 18 and 21, can actually occur. In this case, all sub-boxes of the last tessellation of X have been discarded, and this happens exactly for $\widehat{v}' \geq \widetilde{v}$ for all sub-boxes X'. This implies

$$\min_{x \in X'} f(x) = v' \geq \widehat{v}' \geq \widetilde{v} = f(\widetilde{x})$$

for each sub-box X' and thus $\min_{x \in X} f(x) \geq f(\widetilde{x})$. Therefore, list $= \emptyset$ holds if and only if $\widetilde{x}$ has exactly identified a global minimal point. This case can occur, for example, when P possesses a global minimal point in a vertex of X and if this vertex is also identified as an optimal point of a relaxation on a sub-box.

Algorithm 3.3: Global minimization of a box-constrained function via αBB

Input: $X = [\underline{x}, \overline{x}] \in \mathbb{R}^n$, $f \in C^2(X, \mathbb{R})$ with factorizable Hessian matrix $D^2 f$, termination tolerance $\varepsilon > 0$

Output: ε-optimal feasible point $\tilde{x}$ of f on X, i.e. $\tilde{x} \in X$ with $v \leq f(\tilde{x}) \leq v + \varepsilon$

1 **begin**
2   Compute some $\alpha \geq \max\{0, -\min_{x \in X} \lambda_{\min}(x)\}$.
3   Set $\tilde{x} = m(X)$ and $\tilde{v} = f(\tilde{x})$.
4   Set $X^\star = X$, $\widehat{v}^\star = -\infty$ and list $= (X^\star, \widehat{v}^\star)$.
5   **repeat**
6   Halve $X^\star$ along a longest edge and name the new boxes X^1, X^2, i.e., choose some $i \in \{1, \ldots, n\}$ with

$$\overline{x}_i^\star - \underline{x}_i^\star = \max_{j=1,\ldots,n} \left(\overline{x}_j^\star - \underline{x}_j^\star \right)$$

        and set

$$X^1 = X_1^\star \times \ldots \times [\underline{x}_i^\star, m(X_i^\star)] \times \ldots \times X_n^\star$$

        and

$$X^2 = X_1^\star \times \ldots \times [m(X_i^\star), \overline{x}_i^\star] \times \ldots \times X_n^\star.$$

7   Remove $(X^\star, \widehat{v}^\star)$ from list.
8   **for** $\ell = 1, 2$ **do**
9   Calculate a minimal point $\widehat{x}^\ell$ and the minimal value $\widehat{v}^\ell$ of $\widehat{f}_\alpha^\ell(x) = f(x) + \frac{\alpha}{2}\left(\underline{x}^\ell - x\right)^\intercal \left(\overline{x}^\ell - x\right)$ on X^ℓ (e.g., using a method from Sect. 2.8).
10   **if** $\widehat{v}^\ell < \tilde{v}$ **then**
11   Add the pair $\left(X^\ell, \widehat{v}^\ell\right)$ to list.
12   **if** $f(\widehat{x}^\ell) < \tilde{v}$ **then**
13   Set $\tilde{x} = \widehat{x}^\ell$ and $\tilde{v} = f(\widehat{x}^\ell)$.
14   Remove all pairs $(X', \widehat{v}')$ with $\widehat{v}' \geq \tilde{v}$ from list.
15   **end**
16   **end**
17   **end**
18   **if** list $\neq \emptyset$ **then**
19   Select some $(X^\star, \widehat{v}^\star)$ with minimal $\widehat{v}^\star$ from list.
20   **end**
21   **until** list $= \emptyset$ **or** $\tilde{v} - \widehat{v}^\star \leq \varepsilon$
22 **end**

In contrast to Algorithm 3.2, it is initially unclear whether the termination criterion in line 21 of Algorithm 3.3 is always met after a finite number of steps. The following theorem answers this question. It requires a quantification of the

3.6 Branch-and-Bound for Box-Constrained Problems

statement that the sub-boxes become smaller when descending into the branching tree. For this purpose, we number the levels of the tree that arise from successive branching, starting with the root as level 0.

Lemma 3.6.1 *If the box $X \in \mathbb{R}^n$ is successively subdivided by halving along longest edges, then each box from level N of the branching tree has at most the width $\left(1 - \frac{3}{4n}\right)^{\frac{N}{2}} \cdot w(X)$.*

Proof Let Y be a box and Z one of the two boxes resulting from Y according to the halving rule from line 6 in Algorithm 3.3. If i denotes the index of a longest edge, then the ratio of the squared diagonal lengths satisfies

$$\frac{\|\overline{z} - \underline{z}\|_2^2}{\|\overline{y} - \underline{y}\|_2^2} = \frac{\sum_{j=1}^n (\overline{z}_j - \underline{z}_j)^2}{\|\overline{y} - \underline{y}\|_2^2} = \frac{\sum_{j=1}^n (\overline{y}_j - \underline{y}_j)^2 - (\overline{y}_i - \underline{y}_i)^2 + \left(\frac{\overline{y}_i - \underline{y}_i}{2}\right)^2}{\|\overline{y} - \underline{y}\|_2^2}$$

$$= 1 - \frac{3}{4} \frac{(\overline{y}_i - \underline{y}_i)^2}{\|\overline{y} - \underline{y}\|_2^2} \leq 1 - \frac{3}{4n},$$

where the inequality is due to the maximum property of $\overline{y}_i - \underline{y}_i$ and

$$\|\overline{y} - \underline{y}\|_2^2 = \sum_{j=1}^n (\overline{y}_j - \underline{y}_j)^2 \leq n(\overline{y}_i - \underline{y}_i)^2.$$

When moving to the next deeper level, the diagonal lengths are thus reduced by at least the factor $\sqrt{1 - \frac{3}{4n}}$. This proves the assertion. □

For example, the reduction factor for $n = 2$ is less than $\sqrt{\frac{5}{8}} \approx 0.79$ and for $n = 3$ under $\frac{\sqrt{3}}{2} \approx 0.89$. The terms $\sqrt{1 - \frac{3}{4n}}$ tend to 1 for $n \to \infty$, which quantifies the inevitable slowing effect of high dimensions on the convergence of the method.

Theorem 3.6.2 *Algorithm 3.3 terminates after a finite number of steps.*

Proof We show that after a finite number of iterations the termination criterion in line 21 applies. If after a finite number of iterations list = ∅ holds, this is certainly the case. In the following we may therefore assume list ≠ ∅ for all considered iterations.

In each iteration, a box X^* with associated bound $\widehat{v}^*$ is therefore selected from the list in line 19. X^* must have been added to the list in this or an earlier iteration in line 11, after which in lines 12 and 13 the value $f(\widehat{x}^*)$ contributed to the determination of $\widetilde{v}$. In the current iteration, therefore, $\widetilde{v} \leq f(\widehat{x}^*)$ holds, and thus in line 21

$$\widetilde{v} - \widehat{v}^* \leq f(\widehat{x}^*) - \widehat{v}^* \leq \frac{\alpha}{8} w(X^*)^2.$$

In order for $\widetilde{v} - \widehat{v}^*$ to fall below ε after a finite number of steps as required, it is sufficient to show that the width of the box X^* selected in line 19 is sufficiently small after a sufficient number of iterations. If X^* is at level N of the branching tree, then by Lemma 3.6.1

$$w(X^*) \leq \left(1 - \frac{3}{4n}\right)^{\frac{N}{2}} w(X)$$

holds, and thus

$$\widetilde{v} - \widehat{v}^* \leq \frac{\alpha}{8} \left(1 - \frac{3}{4n}\right)^N w(X)^2.$$

The tolerance $\widetilde{v} - \widehat{v}^* \leq \varepsilon$ is therefore guaranteed if $\frac{\alpha}{8}(1 - \frac{3}{4n})^N w(X)^2 \leq \varepsilon$ holds, that is, for

$$N \geq \frac{\log\left(\frac{8\varepsilon}{\alpha w(X)^2}\right)}{\log\left(1 - \frac{3}{4n}\right)}.$$

The latter occurs for the first time at level

$$N = \left\lceil \frac{\log\left(\frac{8\varepsilon}{\alpha w(X)^2}\right)}{\log\left(1 - \frac{3}{4n}\right)} \right\rceil^+ \tag{3.4}$$

where, analogously to our previous observation for uniformly refined tessellations, a sufficiently coarse tolerance ε leads to $N = 0$, i.e., termination in the root node.

The question remains whether for $N \geq 1$ the algorithm is guaranteed to reach level N of the branching tree after a finite number of iterations. In the best case, one reaches this level after $N + 1$ iterations (counting the root node), if a deeper level is reached in every iteration. In the worst case, however, the method first constructs all sub-boxes at level $N - 1$ before it switches to level N. This requires

$$1 + 2 + \ldots + 2^{N-1} = \frac{2^N - 1}{2 - 1} = 2^N - 1$$

3.6 Branch-and-Bound for Box-Constrained Problems

iterations. But at the latest in iteration 2^N the method then reaches level N of the tree. □

The next result follows immediately from this proof.

> **Corollary 3.6.3** *With the value N from (3.3), Algorithm 3.3 requires at most $N + 1$ steps in the best case, and at most 2^N steps in the worst case.*

The statement in Corollary 3.6.3 about the best case is not particularly helpful, as one does not know a priori whether a good or a bad case is on hand. In practice, however, one rarely observes run times of the order 2^N.

Example 3.6.4 For the data $n = 3$, $w(X) = 2\sqrt{3}$, $\alpha = 1$ and $\varepsilon = 10^{-3}$ from Example 3.5.6 one obtains $N = 26$ and $2^N = 67.108.864$.

We conclude this section with some remarks on Algorithm 3.3:

- If a point $\widetilde{x} \in X$ with a lower objective function value than $f(m(X))$ is already known before applying the method, for example as an observed initial scenario or as the result of some heuristic, line 3 can be replaced by the initialization $\widetilde{v} = f(\widetilde{x})$. The discarding steps then may be expected to kick in earlier, so that the list is kept shorter, and also the value $\widetilde{v} - \widehat{v}^\star$ is smaller from the beginning. Both can lead to a faster termination of the procedure.
- Recalculating the α-values on each sub-box can lead to significantly better bounds and thus to a significant reduction in the number of iterations. However, there is a trade-off to the effort of the α-calculations, i.e., the CPU time may possibly increase. The better alternative is problem dependent.
- The 'cleanup step' in line 14 can take a lot of time with long lists. Then it makes sense to perform it not in every iteration, but only sporadically.
- Since, in graph terminology, the boxes from the list form nodes of the branching tree, the choice of the box in line 19 is also called *node selection*.
- Even if the procedure is terminated prematurely, useful information is generated, namely *valid* upper and lower bounds on v as well as a rough localization of global minimal points as a result of discarding sub-boxes. Also, if required, a known starting point can at least be replaced by a new 'best known point' $\widetilde{x}$.
- By adjusting some of the inequalities, it is not difficult to modify the procedure so that the boxes in the list at termination cover all global minimal points.
- The relaxation of f by the αBB technique can be replaced by other approaches, provided estimates for the maximal error per box size are known (e.g., as in Sect. 3.9). For some nonconvex functions, even explicit formulas for the envelope functions are known [10].

3.7 Branch-and-Bound for Convexly Constrained Problems

In this section, we discuss problems that are somewhat more elaborate to handle from the perspective of the αBB method than those from Sect. 3.6, namely problems with only convex instead of box-shaped feasible sets. For this, we consider a C^2-problem

$$P: \quad \min f(x) \quad \text{s.t.} \quad x \in M$$

with convexly described feasible set

$$M = \{x \in X \mid g_i(x) \leq 0, \ i \in I, \ h_j(x) = 0, \ j \in J\}$$

and $X \in \mathbb{R}^n$, $|I| < \infty$ and $|J| < n$. Nonconvexity of P can then only be due to the nonconvexity of f.

We will apply the box subdivision strategy from Sect. 3.6 to the box X and thus divide the set M into subsets of the form

$$M^\ell = M \cap X^\ell = \{x \in X^\ell \mid g_i(x) \leq 0, \ i \in I, \ h_j(x) = 0, \ j \in J\}.$$

Almost all results and remarks from the purely box-constrained case then transfer to convexly constrained problems. However, two new effects occur, which we will discuss next.

Convexly Constrained Subproblems

A first difference to the box-constrained case is that in the convex relaxation $\widehat{P}^\ell$ of P on X^ℓ the function $\widehat{f}_\alpha^\ell$ is not box-constrained to X^ℓ, but convexly constrained to the set M^ℓ (Definition 3.2.4a). Since M^ℓ is again convex, it does not need to be relaxed to a set $\widehat{M}^\ell$. Every optimal point of

$$\widehat{P}^\ell: \quad \min \widehat{f}_\alpha^\ell(x) \quad \text{s.t.} \quad x \in M^\ell$$

therefore also lies in M and can thus be used to improve the upper bound $\widetilde{v}$ of v.

Inconsistent Subproblems

A more significant difference to the box-constrained case is that sets M^ℓ may be empty, as Fig. 3.22 shows. A relaxed subproblem $\widehat{P}^\ell$ can therefore be unsolvable due to inconsistency, and the associated sub-box X^ℓ then needs to be discarded. Since $\widehat{P}^\ell$ is a convex optimization problem, if a nonlinear optimization method for $\widehat{P}^\ell$ reports that it finds no feasible points, one can trust that indeed $M^\ell = \emptyset$ holds, and thus set $\widehat{v}^\ell = +\infty$ as usual.

Also, the initialization of $\widetilde{v}$ by evaluating f at $m(X)$ can fail, when $m(X)$ does not lie in M. If no alternative feasible starting point is known, one may set $\widetilde{v} = +\infty$ and

Fig. 3.22 Empty subset M^ℓ of the feasible set M

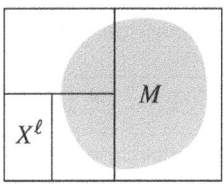

let Algorithm 3.4 generate feasible points and thus finite values for $\widetilde{v}$. If M is not empty, this happens for at least one of the first two sub-boxes of X.

If, on the other hand, M is empty, the infimum $\widehat{v} = +\infty$ is assigned to each of the first two sub-boxes of X. In addition, $\widetilde{v}$ must also have been initialized to $+\infty$, so that Algorithm 3.4 terminates with the report of unsolvability.

Otherwise, in complete analogy to Theorem 3.6.2, one proves the following theorem.

Theorem 3.7.1 *Algorithm 3.4 terminates after a finite number of steps.*

Corollary 3.6.3 and the above remarks on Algorithm 3.3 apply accordingly. The requirement of twice continuously differentiable functions g_i, $i \in I$, is not essential here and can be weakened to a smoothness requirement under which the optimization problems in line 9 and line 12 can be algorithmically treated (e.g., with a method from Sect. 2.8).

3.8 Branch-and-Bound for Nonconvex Problems

Finally, in this section, we deal with general optimization problems where, in addition to the objective function, also the feasible set may be nonconvex. For this, we consider a C^2-problem

$$P: \quad \min f(x) \quad \text{s.t.} \quad x \in M$$

with feasible set

$$M = \{x \in X | \ g_i(x) \le 0, \ i \in I, \ h_j(x) = 0, \ j \in J\}$$

and $X \in \mathbb{R}^n$, $|I| < \infty$ and $|J| < n$. In the simplest case, the nonconvexity arises because only *one* of the involved functions violates the convexity assumptions:

- The objective function f is not convex (or it is not known whether f is convex).
- For some $i \in I$, the inequality constraint function g_i is not convex.
- For some $j \in J$, the equality constraint function h_j is not linear.

Algorithm 3.4: Global minimization of a convexly constrained function via αBB

Input: $X = [\underline{x}, \overline{x}] \in \mathbb{R}^n$, $f \in C^2(X, \mathbb{R})$ with factorizable Hessian matrix $D^2 f$, convex $g_i \in C^2(X, \mathbb{R})$, $i \in I$, linear h_j, $j \in J$, termination tolerance $\varepsilon > 0$

Output: ε-optimal feasible point $\widetilde{x}$ of P, i.e. $\widetilde{x} \in M$ with $v \leq f(\widetilde{x}) \leq v + \varepsilon$, or report of unsolvability

1 **begin**
2   Compute some $\alpha \geq \max\{0, -\min_{x \in X} \lambda_{\min}(x)\}$.
3   **if** $\widetilde{x} \in M$ *is known* **then** set $\widetilde{v} = f(\widetilde{x})$ **else** set $\widetilde{v} = +\infty$.
4   Set $X^\star = X$, $\widehat{v}^\star = -\infty$ and list $= (X^\star, \widehat{v}^\star)$.
5   **repeat**
6   Halve $X^\star$ along a longest edge and name the new boxes X^1, X^2.
7   Remove $(X^\star, \widehat{v}^\star)$ from list.
8   **for** $\ell = 1, 2$ **do**
9   Compute the infimum $\widehat{v}^\ell$ of $\widehat{f}_\alpha^\ell(x)$ on $M^\ell = M \cap X^\ell$.
10   **if** $\widehat{v}^\ell < \widetilde{v}$ **then**
11   Add the pair $(X^\ell, \widehat{v}^\ell)$ to list.
12   Compute a minimal point $\widehat{x}^\ell$ of $\widehat{f}_\alpha^\ell(x)$ on M^ℓ.
13   **if** $f(\widehat{x}^\ell) < \widetilde{v}$ **then**
14   Set $\widetilde{x} = \widehat{x}^\ell$ and $\widetilde{v} = f(\widehat{x}^\ell)$.
15   Remove all pairs $(X', \widehat{v}')$ with $\widehat{v}' \geq \widetilde{v}$ from list.
16   **end**
17   **end**
18   **end**
19   **if** *list* $\neq \emptyset$ **then**
20   Select some $(X^\star, \widehat{v}^\star)$ with minimal $\widehat{v}^\star$ from list.
21   **end**
22   **until** list $= \emptyset$ *or* $\widetilde{v} - \widehat{v}^\star \leq \varepsilon$
23 **end**
24 **case** $\widetilde{v} < +\infty$
25     $\widetilde{x}$ is an ε-optimal feasible point of P.
26 **case** $\widetilde{v} = +\infty$
27   P is unsolvable due to inconsistency.

In many applications several such violations occur simultaneously.

For the branch-and-bound approach, a convex relaxation $\widehat{P}$ of P is again needed, where of course only those functions are modified for which convexity or linearity is not already clear (cf. the corresponding treatment of the sets K and L in Algorithm 2.1):

- If f is not convex, we calculate some $\alpha \geq \max\{0, -\min_{x \in X} \lambda_f(x)\}$, where $\lambda_f(x)$ denotes the smallest eigenvalue of $D^2 f(x)$.
- If g_i is not convex for some $i \in I$, we calculate some $\beta_i \geq \max\{0, -\min_{x \in X} \lambda_{g_i}(x)\}$.
- If h_j is not linear for some $j \in J$, we decompose the equality constraint $h_j(x) = 0$ into the two inequality constraints $\pm h_j(x) \leq 0$ and relax these if necessary,

3.8 Branch-and-Bound for Nonconvex Problems

i.e., we calculate some $\gamma_j^+ \geq \max\{0, -\min_{x \in X} \lambda_{h_j}(x)\}$, if h_j is not convex, and some $\gamma_j^- \geq \max\{0, -\min_{x \in X} \lambda_{-h_j}(x)\}$, if $-h_j$ is not convex.

The most general case to treat nonlinear equality constraints is illustrated by the function $h(x) = x_2 - \sin(x_1)$ on $X = [0, 2\pi] \times [-1, 1]$. Neither h nor $-h$ is convex, so both h and $-h$ need to be convexly relaxed. On the other hand, for the function $h(x) = 1 - x_1^2 - x_2^2$, only h is not convex, but $-h$ is. Therefore, only h needs to be convexly relaxed, while the inequality $-h(x) \leq 0$ can be copied to the relaxed optimization problem without modification.

The convex relaxation $\widehat{P}$ of P on X with feasible set $\widehat{M}$ is constructed using these required convex relaxations of the defining functions of P. If the box X is replaced again by a sub-box X^ℓ, the constraints of the feasible set $\widehat{M}^\ell$ of the sub-problem $\widehat{P}^\ell$ arise accordingly.

Example 3.8.1 We consider the problem

$$P: \quad \min f(x) \quad \text{s.t.} \quad g_1(x) \leq 0, \ g_2(x) \leq 0, \ x \in X = [\underline{x}, \overline{x}],$$
$$h_1(x) = 0, \ h_2(x) = 0, \ h_3(x) = 0$$

with f, g_1 convex, g_2 nonconvex, h_1 linear, h_2 nonlinear convex and h_3 neither convex nor concave. Then one needs to calculate β_2, γ_2^-, γ_3^+ as well as γ_3^-, and obtains

$$\widehat{P}: \quad \min f(x) \quad \text{s.t.} \quad x \in X, \ g_1(x) \leq 0,$$
$$g_2(x) + \frac{\beta_2}{2}(\underline{x} - x)^\mathsf{T}(\overline{x} - x) \leq 0,$$
$$h_1(x) = 0,$$
$$h_2(x) \leq 0,$$
$$-h_2(x) + \frac{\gamma_2^-}{2}(\underline{x} - x)^\mathsf{T}(\overline{x} - x) \leq 0,$$
$$h_3(x) + \frac{\gamma_3^+}{2}(\underline{x} - x)^\mathsf{T}(\overline{x} - x) \leq 0,$$
$$-h_3(x) + \frac{\gamma_3^-}{2}(\underline{x} - x)^\mathsf{T}(\overline{x} - x) \leq 0.$$

For the branch-and-bound method, compared to the convexly constrained case, two further new effects occur in the case of a nonconvex feasible set of P.

Fig. 3.23 Empty feasible set $\widehat{M}^\ell$

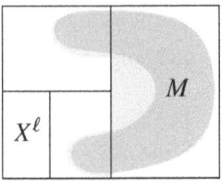

Fig. 3.24 $M^\ell = \emptyset$ does not imply $\widehat{M}^\ell = \emptyset$

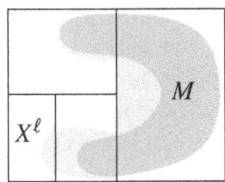

Inconsistency of Subproblems Potentially not Immediately Detectable

As in the convexly constrained case, inconsistent relaxed subproblems $\widehat{P}^\ell$ can occur, as Fig. 3.23 illustrates for a nonconvex set M. Since $\widehat{M}^\ell$ is a convex set, its inconsistency can be checked algorithmically. According to Theorem 3.2.3a, $\widehat{M}^\ell \supseteq M^\ell = M \cap X^\ell$ holds. From $\widehat{M}^\ell = \emptyset$ follows $M^\ell = \emptyset$ (Fig. 3.23), and the sub-box X^ℓ can then be discarded.

Conversely, $M^\ell = \emptyset$ does *not* imply that also $\widehat{M}^\ell = \emptyset$ holds, as Fig. 3.24 illustrates. The box X^ℓ is then not discarded despite $M^\ell = \emptyset$ and remains in the list. Fortunately, finitely many further subdivisions of X^ℓ can identify the inconsistency of all corresponding subsets of M^ℓ, as the following result shows.

Theorem 3.8.2 *Given a sub-box X^ℓ, let $M^\ell = M \cap X^\ell = \emptyset$. Then after a finite number of subdivisions $\widehat{M}^k = \emptyset$ holds for every sub-box X^k of X^ℓ.*

Proof We conduct a proof by contradiction and assume that there is an infinite sequence of sub-boxes $X^k \subseteq X^\ell$ with $\widehat{M}^k \neq \emptyset$. Then $\lim_k w(X^k) = 0$ holds, and one can (possibly after choosing a subsequence) assume that $X^{k+1} \subseteq X^k$, $k \in \mathbb{N}$, holds. The box center points $m(X^k)$ thus possess a limit point $x^* \in X^\ell$. Obviously, $x^* \in X^k$ also holds for all $k \in \mathbb{N}$.

Because of $M^\ell = \emptyset$, at least one constraint must be violated at x^*, say $g_i(x^*) > 0$ with some $i \in I$. The continuity of g_i implies that for all sufficiently large $k \in \mathbb{N}$ with $x^* \in X^k$ also all other elements of X^k fulfill this inequality. The Weierstrass theorem thus implies $\min_{x \in X^k} g_i(x) = c > 0$. According to Lemma 3.4.2c, for sufficiently large $k \in \mathbb{N}$ every $x \in X^k$ therefore fulfills

$$(\widehat{g}_i)_{\beta_i}^k(x) \geq g_i(x) - \frac{\beta_i}{8} w(X^k)^2 \geq c - \frac{\beta_i}{8} w(X^k)^2 \geq \frac{c}{2} > 0.$$

This implies $\widehat{M}^k = \emptyset$, contradicting the assumption. □

Possible Infeasibility of Minimal Points of the Relaxations

A *significant* problem with nonconvexly constrained problems is that an optimal point $\widehat{x}^\ell$ of $\widehat{P}^\ell$ does not necessarily have to be feasible for P since, for example,

3.8 Branch-and-Bound for Nonconvex Problems

boundary points of $\widehat{M}^\ell$, where minimal points love to reside, do not necessarily also lie in M^ℓ (Fig. 3.23).

This has two consequences. Firstly, as with each outer approximation method, one can only expect that the points $\tilde{x}$ generated by the algorithm are *asymptotically* feasible, that is, the approximation of an optimal point generated after finitely many steps is not necessarily an element of M. As is common with outer approximation methods, this *could* be handled by defining a 'maximal allowed infeasibility' of $\tilde{x}$ through a further tolerance $\varepsilon_M > 0$ and a penalty function for M, for example

$$\rho(\tilde{x}) := \sum_{i \in I} g_i^+(\tilde{x}) + \sum_{j \in J} |h_j(\tilde{x})| \leq \varepsilon_M,$$

where $g_i^+(\tilde{x}) = \max\{0, g_i(\tilde{x})\}$ again denotes the positive part of $g_i(\tilde{x})$. With a similar argument as in the proof of Theorem 3.8.2 it can be shown that for any $\varepsilon_M > 0$ this ε_M-feasibility is fulfilled after finitely many box subdivisions.

The second consequence, however, is far-reaching and cannot be handled by this construction: For $\widehat{x}^\ell \notin M$ the value $f(\widehat{x}^\ell)$ is not necessarily an upper bound for v and therefore cannot be used to update $\tilde{v}$ (Fig. 3.25). As a way out, one first checks whether $\widehat{x}^\ell \in M$ happens to hold, just by evaluating the functions g_i, $i \in I$, h_j, $j \in J$, at $\widehat{x}^\ell$. If so, $f(\widehat{x}^\ell)$ is used to update the upper bound $\tilde{v}$ of v. Otherwise, various *heuristics* for further action are common:

- One can hope that the algorithm will eventually determine feasible points on its own. Unfortunately, this is not guaranteed.
- One can try to determine a local minimal point x^ℓ of f on $M^\ell = M \cap X^\ell$ using a method of nonlinear optimization and use the value $f(x^\ell)$ to update $\tilde{v}$. In addition to the effort of this approach, drawbacks are that an element of a nonconvexly described set M^ℓ is not necessarily found algorithmically, even if it exists, and that it is unclear whether the termination criterion $\tilde{v} - \tilde{v}^* \leq \varepsilon$ will be met after a finite number of steps.
- One allows ε_M-feasible points for the update of $\tilde{v}$, i.e., $\widehat{x}^\ell$ with $\rho(\widehat{x}^\ell) < \varepsilon_M$, hoping that this results in only a small violation of the upper bounding property of $\tilde{v}$. However, Fig. 3.26 illustrates that even here arbitrarily wrong values of $\tilde{v}$

Fig. 3.25 An infeasible point does not provide an upper bound for v

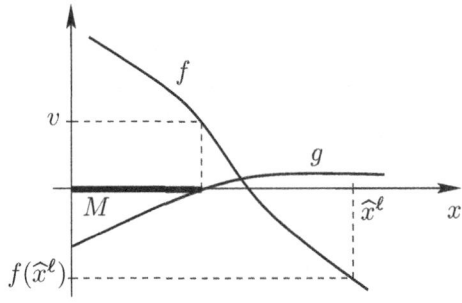

Fig. 3.26 An ε_M-feasible point does not provide an upper bound for v

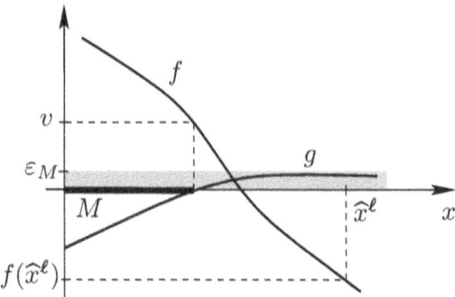

can arise. An approach to estimate the resulting error using Lipschitz constants will be discussed in Sect. 3.9. However, neither this will yield a remedy.

For a *deterministic* approach to the guaranteed determination of a feasible point in the case $\widehat{x}^\ell \notin M$, see [26]. If this technique is used in line 14 of Algorithm 3.5, similar remarks apply as to Algorithm 3.3, and it is again possible to show the termination of Algorithm 3.5 after a finite number of steps, as well as complexity results. The arising convex subproblems can be solved, for example, by the methods from Sect. 2.8.

Exercise 3.8.3 According to Theorem 3.2.5c, for the optimal value of the convex hull problem of an optimization problem with a nonconvex feasible set, $\widehat{\widehat{v}} < v$ can hold. In such a case, the optimal values $\widehat{v}$ of all convexly relaxed problems retain a positive distance from v because $\widehat{v} \leq \widehat{\widehat{v}} < v$. Why does this not contradict the fact that a convergence result can still be shown for the iteratively refined relaxations from Algorithm 3.5?

3.9 Lipschitz Properties

This concluding section discusses some useful properties of Lipschitz continuous functions for global optimization. After an introduction to Lipschitz continuity in Sect. 3.9.1, in Sect. 3.9.2 we try to use Lipschitz estimates to construct the hard to access upper bounds for the optimal value in Algorithm 3.5. Since this will turn out to be only moderately successful, Sect. 3.9.3 instead shows how the idea of the αBB relaxation can be significantly modified when Lipschitz constants are available.

3.9 Lipschitz Properties

Algorithm 3.5: Global minimization of a nonconvex problem via αBB

Input: $X = [\underline{x}, \overline{x}] \in \mathbb{R}^n$, $f, g_i, h_j \in C^2(X, \mathbb{R})$, $i \in I$, $j \in J$, with
- f convex or $D^2 f$ factorizable,
- g_i convex or $D^2 g_i$ factorizable, $i \in I$,
- h_j linear or
    - h_j convex or $D^2 h_j$ factorizable, $j \in J$,
    - $-h_j$ convex or $-D^2 h_j$ factorizable, $j \in J$,
  termination tolerance $\varepsilon > 0$

Output: ε-optimal feasible point $\tilde{x}$ of P, i.e. $\tilde{x} \in M$ with $v \leq f(\tilde{x}) \leq v + \varepsilon$, or report of unsolvability (with line 14 e.g. according to [26])

1 **begin**
2   Calculate the required $\alpha, \beta_i, i \in I, \gamma_j^{\pm}, j \in J$.
3   **if** $\tilde{x} \in M$ *is known* **then** set $\tilde{v} = f(\tilde{x})$ **else** set $\tilde{v} = +\infty$.
4   Set $X^\star = X, \hat{v}^\star = -\infty$ and list $= (X^\star, \hat{v}^\star)$.
5   **repeat**
6   Halve $X^\star$ along a longest edge and call the new boxes X^1, X^2.
7   Remove $(X^\star, \hat{v}^\star)$ from list.
8   **for** $\ell = 1, 2$ **do**
9   Calculate the infimum $\hat{v}^\ell$ of $\widehat{f}_\alpha^\ell(x)$ on $\widehat{M}^\ell$.
10   **if** $\hat{v}^\ell < \tilde{v}$ **then**
11   Add the pair $(X^\ell, \hat{v}^\ell)$ to list.
12   Calculate a minimal point $\hat{x}^\ell$ of $\widehat{f}_\alpha^\ell(x)$ on $\widehat{M}^\ell$.
13   **if** $\hat{x}^\ell \notin M$ **then**
14   Try to replace $\hat{x}^\ell$ with a feasible point.
15   **end**
16   **if** $\hat{x}^\ell \in M$ *and* $f(\hat{x}^\ell) < \tilde{v}$ **then**
17   Set $\tilde{x} = \hat{x}^\ell$ and $\tilde{v} = f(\hat{x}^\ell)$.
18   Remove all pairs $(X', \hat{v}')$ with $\hat{v}' \geq \tilde{v}$ from list.
19   **end**
20   **end**
21   **end**
22   **if** *list* $\neq \emptyset$ **then**
23   Select some $(X^\star, \hat{v}^\star)$ with minimal $\hat{v}^\star$ from list.
24   **end**
25   **until** *list* $= \emptyset$ *or* $\tilde{v} - \hat{v}^\star \leq \varepsilon$
26 **end**
27 **case** $\tilde{v} < +\infty$
28     $\tilde{x}$ is an ε-optimal feasible point of P.
29 **case** $\tilde{v} = +\infty$
30   P is unsolvable due to inconsistency.

3.9.1 Properties of Lipschitz Continuous Functions

Definition 3.9.1 (Lipschitz Continuity) For $X \subseteq \mathbb{R}^n$, a function $f : X \to \mathbb{R}$ is called *Lipschitz continuous*, if there exists a constant $L > 0$ with

$$\forall x, y \in X : \quad |f(x) - f(y)| \leq L \cdot \|x - y\|_2.$$

L is then called a *Lipschitz constant* for f on X.

Lipschitz continuity can also be considered with respect to any other norm $\|\cdot\|$, instead of $\|\cdot\|_2$.

For $x = y$, the Lipschitz condition is not interesting. For $x \neq y$, it states that the secant through the points $(x, f(x))$ and $(y, f(y))$ on the graph of f has a 'bounded slope'. The appearance of the absolute value is explained by the fact that for $n > 1$ not all arguments x and y are comparable by $\leq$, and therefore it is not clear whether to measure the secant slope from x in the direction of y or in the opposite direction. In the absolute value expression $|f(x) - f(y)| / \|x - y\|_2$ this does not matter. For a Lipschitz continuous function, this expression is therefore bounded by the same constant $L > 0$ for any choice of $x, y \in X$, and the 'variation' of f is in this sense bounded. The following examples illustrate the concept of Lipschitz continuity.

- The function $f(x) = \sqrt[3]{x}$ is not Lipschitz continuous on $X = [-1, 1]$. Figure 3.27 illustrates why this is geometrically clear: With the choices $x = 0$ and $y^k = 1/k$, for $k \to \infty$ arbitrarily steep secants can be generated on the graph of f.
- The function $f(x) = x^2$ is not Lipschitz continuous on $X = \mathbb{R}$. This is also geometrically clear, and formally it can be seen as follows: For all $x, y \in \mathbb{R}$ we have

$$|f(x) - f(y)| = |x^2 - y^2| = |x + y| \cdot |x - y|.$$

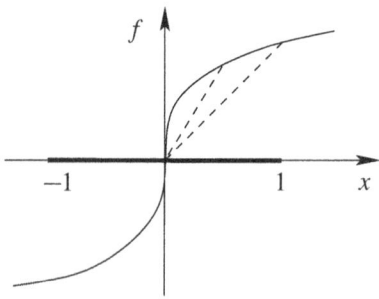

Fig. 3.27 Lack of Lipschitz continuity

3.9 Lipschitz Properties

Since the expression $|x + y|$ becomes arbitrarily large with suitable choices of $x, y \in \mathbb{R}$, no Lipschitz constant L can be found. More generally, with this argument it can be seen that $f(x) = x^2$ is Lipschitz continuous on every bounded set X, and not on any unbounded set X.

- The function $f(x) = |x|$ is not differentiable on $\mathbb{R}$, but it is convex and Lipschitz continuous.
- The function $f(x) = -|x|$ is neither differentiable nor convex on $\mathbb{R}$, but it is Lipschitz continuous.

The last two examples show that Lipschitz continuous functions are not necessarily differentiable. However, it is easy to prove that they are at least continuous. On the other hand, a profound result from calculus (Rademacher's theorem) states that Lipschitz continuous functions are in a certain sense 'almost everywhere' differentiable. This can be used, for example, to extend the idea of the convex subdifferential to Lipschitz continuous functions [6].

For some Lipschitz constant L of f on X, every $L' > L$ is also a Lipschitz constant of f on X. In general, one is interested in identifying a possibly small or even the smallest Lipschitz constant, as this provides the highest amount of information about the behavior of f on X. However, if this smallest L is too cumbersome to calculate, one can also be satisfied with larger Lipschitz constants, at the price of them providing coarser descriptions of the behavior of f on X. The following result provides a way to calculate Lipschitz constants.

Lemma 3.9.2

(a) Let $X \subseteq \mathbb{R}^n$ be a nonempty and convex set, let the function $f : X \to \mathbb{R}$ be differentiable, and assume

$$\sup_{x \in X} \|\nabla f(x)\|_2 < +\infty.$$

Then f is Lipschitz continuous on X, and every $L > 0$ with

$$L \geq \sup_{x \in X} \|\nabla f(x)\|_2$$

is a Lipschitz constant of f on X.

(b) Let $X \subseteq \mathbb{R}^n$ be a nonempty, convex and compact set, and let the function $f : X \to \mathbb{R}$ be continuously differentiable. Then f is Lipschitz continuous on X, and every $L > 0$ with

$$L \geq \max_{x \in X} \|\nabla f(x)\|_2$$

is a Lipschitz constant of f on X.

Proof Let $x, y \in X$. Since the Lipschitz condition for $x = y$ is clear, let $x \neq y$. By the mean value theorem, there exists some z on the connecting line between x and y with

$$f(x) = f(y) + \langle \nabla f(z), x - y \rangle,$$

where the convexity of X also implies $z \in X$. The Cauchy-Schwarz inequality implies

$$|f(x) - f(y)| = |\langle \nabla f(z), x - y \rangle| \leq \|\nabla f(z)\|_2 \cdot \|x - y\|_2$$

$$\leq \left(\sup_{z \in X} \|\nabla f(z)\|_2 \right) \cdot \|x - y\|_2 .$$

The independence of the number $\sup_{z \in X} \|\nabla f(z)\|_2$ from x and y yields the assertion of statement a. The assertion of statement b follows from statement a, because under the additional assumptions, according to the Weierstrass theorem, the supremum is attained. □

Exercise 3.9.3 Since one is interested in smallest possible Lipschitz constants, in Lemma 3.9.2b it may seem more appropriate to directly define $L := \max_{x \in X} \|\nabla f(x)\|_2$. In which (usually uninteresting) case would then a formal problem arise?

Exercise 3.9.4 Show, using the set $X = \{0\} \times \mathbb{R}$ and the function $f(x) = x_1$, that it may be possible to improve the Lipschitz constant from Lemma 3.9.2a.

Exercise 3.9.5 Show, under the conditions of Lemma 3.9.2b, that every $L > 0$ with $L \geq \max_{x \in X} \|\nabla f(x)\|_1$ is a Lipschitz constant for f on X with respect to the ℓ_∞-norm.

From Lemma 3.9.2 we obtain the following result.

Theorem 3.9.6 *Let $X \in \mathbb{R}^n$, $f \in C^1(X, \mathbb{R})$, and $g := \nabla f$ be factorizable with natural interval extension G. If NORM denotes the interval extension of the elementary function $\mathrm{norm}(x) := \|x\|_2$, then every $L > 0$ with $L \geq \overline{NORM(G(X))}$ is a Lipschitz constant of f on X.*

Example 3.9.7 On $X = [0, 2]$, let a Lipschitz constant for the function $f(x) = x^2$ to be determined. Graphically, it is easy to see that no secant can have a higher slope than 4 (namely the tangent slope at $x = 2$) and that this slope can be approximated

3.9 Lipschitz Properties

arbitrarily well by secant slopes. Therefore, $L = 4$ is the best possible Lipschitz constant.

According to Theorem 3.9.6, a Lipschitz constant is calculated as follows: We have $g(x) = \nabla f(x) = f'(x) = 2x$ and thus $G(X) = 2X$. This leads to $\text{NORM}(G(X)) = \text{ABS}(2X)$ and

$$\text{NORM}(G(X)) = \text{ABS}(2[0, 2]) = \text{ABS}([0, 4]) = [0, 4].$$

From Theorem 3.9.6 we thus obtain that $\overline{\text{NORM}}(G(X)) = 4$ is a Lipschitz constant. It matches the best Lipschitz constant determined graphically, but due to the dependency effect and in view of Exercise 3.9.4, one cannot generally expect that Theorem 3.9.6 always provides such a *best* Lipschitz constant.

In a next step, valid upper and lower bounds for functions on sets can be determined with the help of Lipschitz constants.

Lemma 3.9.8 *For $X \subseteq \mathbb{R}^n$ and $f : X \to \mathbb{R}$, let L be a Lipschitz constant, and let $y \in X$ be given. Then it holds*

$$\forall x \in X: \quad f(x) \in [f(y) - L\|x - y\|_2, \ f(y) + L\|x - y\|_2].$$

Proof For all $x \in X$ we have

$$f(x) - f(y) \leq |f(x) - f(y)| \leq L\|x - y\|_2$$

and thus $f(x) \leq f(y) + L\|x - y\|_2$. Similarly, from

$$f(y) - f(x) \leq |f(x) - f(y)| \leq L\|x - y\|_2$$

the inequality $f(x) \geq f(y) - L\|x - y\|_2$ follows, and thus overall the assertion. □

Example 3.9.9 In view of Example 3.9.7, $L = 4$ is a Lipschitz constant of $f(x) = x^2$ on $X = [0, 2]$. With $y = 1$ it follows from Lemma 3.9.8

$$\forall x \in [0, 2]: \quad f(x) \in [1 - 4 \cdot |x - 1|, \ 1 + 4 \cdot |x - 1|].$$

Lemma 3.9.8 can also be interpreted as follows: The graph of f, i.e., the set $\{(x, f(x)) | x \in X\}$, lies for every arbitrary $y \in X$ in the set $\{(x, \alpha) \in X \times \mathbb{R} | |\alpha - f(y)| \leq L\|x - y\|_2\}$. Observe that, while for $n = 1$ this is a double cone with apex $(y, f(y))$, for $n > 1$ only the complement of this set is a double cone.

3.9.2 Direct Application to Algorithm 3.5

For the branch-and-bound method in Algorithm 3.5, the knowledge of a Lipschitz constant of f offers the possibility for an update of the upper bound $\tilde{v}$ of v in the case of infeasibility of $\widehat{x}^\ell$ in line 14. For this, the 'wrong' value $f(\widehat{x}^\ell)$ is corrected with the help of the Lipschitz constant of f to a guaranteed, but possibly coarse upper bound. Unfortunately, the determination of a corresponding feasible point is not straightforward, and the convergence of the method cannot be guaranteed with these upper bounds.

Indeed, let L be a Lipschitz constant of f on X, and let

$$M = \{x \in X | \ g_i(x) \leq 0, \ i \in I, \ h_j(x) = 0, \ j \in J\}$$

be nonempty. From Lemma 3.9.8 with $y = \widehat{x}^\ell$ it follows

$$\forall x \in M: \quad v \leq f(x) \leq f(\widehat{x}^\ell) + L\|x - \widehat{x}^\ell\|_2,$$

so for every choice of $x \in M$ the number $f(\widehat{x}^\ell) + L\|x - \widehat{x}^\ell\|_2$ is an upper bound for v, which can be used to improve $\tilde{v}$. The *best* such upper bound is

$$\min_{x \in M} \left(f(\widehat{x}^\ell) + L\|x - \widehat{x}^\ell\|_2 \right) = f(\widehat{x}^\ell) + L \operatorname{dist}(\widehat{x}^\ell, M),$$

but to determine it, the calculation of the distance $\operatorname{dist}(\widehat{x}^\ell, M)$ from $\widehat{x}^\ell$ to M would be required, i.e., the solution of another nonconvex global optimization problem.

As a remedy, in view of $M \neq \emptyset$ in X there exists a point $x \in M$. In the worst case, x has the maximal distance from $\widehat{x}^\ell$ in X, so a valid, but probably coarse upper bound for v is also

$$f(\widehat{x}^\ell) + L \max_{x \in X} \|x - \widehat{x}^\ell\|_2.$$

The maximal value of $\|x - \widehat{x}^\ell\|_2$ over X can be computed, for example, using the fact that a maximal point can be found in a vertex of X (according to the vertex theorem of convex maximization [32, Cor. 32.3.4]), which leads to 2^n candidates. Because of Exercise 1.3.5 and 1.3.2 we even obtain the explicit representation

$$\max_{x \in X} \|x - \widehat{x}^\ell\|_2 = \sqrt{\max_{x \in X} \sum_{i=1}^n (x_i - \widehat{x}_i^\ell)^2}$$

$$= \sqrt{\sum_{i=1}^n \max_{x_i \in [\underline{x}_i, \overline{x}_i]} (x_i - \widehat{x}_i^\ell)^2}$$

3.9 Lipschitz Properties

$$= \sqrt{\sum_{i=1}^{n} \left(\max\{\widehat{x}_i^\ell - \underline{x}_i, \overline{x}_i - \widehat{x}_i^\ell\}\right)^2}.$$

Unfortunately, this possibility to update the upper bound $\widetilde{v}$ cannot be transferred to arbitrary sub-boxes X^ℓ of X, as one would first have to check

$$M^\ell = \{x \in X^\ell | \; g_i(x) \le 0, \; i \in I, \; h_j(x) = 0, \; j \in J\} \ne \emptyset.$$

Therefore, this approach is of little practical interest.

Also, the knowledge of an 'upper bound for infeasibility' of $\widehat{x}^\ell$ in the sense of ε_M-feasibility,

$$\rho(\widehat{x}^\ell) = \sum_{i \in I} g_i^+(\widehat{x}^\ell) + \sum_{j \in J} |h_j(\widehat{x}^\ell)| \le \varepsilon_M,$$

does not help in a simple way, which we briefly consider for the case $I = \{1\}$ and $J = \emptyset$, i.e. for $\rho(\widehat{x}^\ell) = g^+(\widehat{x}^\ell)$. One might expect that the relaxation $M_{\varepsilon_M} = \{x \in \mathbb{R}^n | \; g(x) \le \varepsilon_M\}$ of the set $M = \{x \in \mathbb{R}^n | \; g(x) \le 0\}$ differs only slightly from M. For example, for every point $\widehat{x}^\ell \in M_{\varepsilon_M}$ one may want to estimate the distance $\text{dist}(\widehat{x}^\ell, M)$ approximately in the form

$$\text{dist}(\widehat{x}^\ell, M) \le \gamma \, \varepsilon_M \qquad (3.5)$$

with a constant $\gamma > 0$ independent of $\widehat{x}^\ell$. Then one would obtain an upper bound $\widetilde{v}$ according to the above considerations through

$$\forall x \in M: \quad v \le f(x) \le f(\widehat{x}^\ell) + L \, \text{dist}(\widehat{x}^\ell, M) \le f(\widehat{x}^\ell) + L \gamma \, \varepsilon_M.$$

Unfortunately, the estimate in (3.4) is related to the 'opposite' of a Lipschitz estimate, namely the estimation of the distance of the arguments by a multiple of the distance of the function values of g^+. A bit more precisely, one sees this as follows: If some $\gamma > 0$ exists, so that for every $\widehat{x}^\ell \in M_{\varepsilon_M}$ there is some $x \in M$ with

$$\|x - \widehat{x}^\ell\|_2 \le \gamma \, |g^+(x) - g^+(\widehat{x}^\ell)|$$

then, due to the feasibility of the $x \in M$ and the non-negativity of $g^+(\widehat{x}^\ell)$

$$\|x - \widehat{x}^\ell\|_2 \le \gamma \, g^+(\widehat{x}^\ell) = \gamma \, \rho(\widehat{x}^\ell) \le \gamma \, \varepsilon_M$$

and thus

$$\text{dist}(\widehat{x}^\ell, M) = \inf_{x \in M} \|x - \widehat{x}^\ell\|_2 \le \gamma \, \varepsilon_M$$

Fig. 3.28 Function ψ in the 'Lipschitz alternative' of Algorithm 3.5

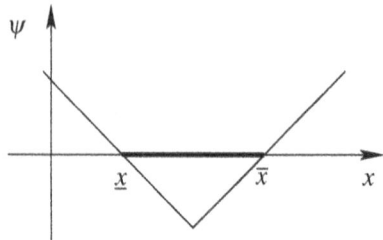

hold. For linear and convex functions, estimates like in (3.4) are indeed possible and known as *Hoffman's lemma* or *global error bounds*, where the constant γ is referred to as a *Hoffman constant* (for details see, e.g., [34]). On the other hand, optimization problems with linear and convex functions can be handled with the techniques from Chap. 2 anyway. For the nonconvex case that interests us here, Fig. 3.26 shows that the specification of error bounds encounters fundamental problems, so that also in dealing with ε_M-admissible points the knowledge of Lipschitz constants hardly helps.

3.9.3 A Variation of Algorithm 3.5

The knowledge of Lipschitz constants also leads to a way to construct relaxations of functions, as an alternative to the αBB technique. These relaxations are not necessarily convex, but still easy to minimize. In the following, we only discuss the case $n = 1$, in which a C^1-function $f : \mathbb{R} \to \mathbb{R}$ is to be minimized on $X = [\underline{x}, \overline{x}] \in \mathbb{IR}$. The function

$$\psi(x) = |x - m(X)| - \frac{w(X)}{2} = \left| x - \frac{\underline{x} + \overline{x}}{2} \right| - \frac{\overline{x} - \underline{x}}{2}$$

takes on the role of $\psi(x) = \frac{1}{2}(\underline{x} - x)(\overline{x} - x)$ from the αBB approach. It is sketched in Fig. 3.28.

Theorem 3.9.10 *Let $X \in \mathbb{IR}$, $f \in C^1(X, \mathbb{R})$ and $L > 0$ be a Lipschitz constant of f on X with $L \geq \max_{x \in X} |f'(x)|$. Then with*

$$\widehat{f}_L(x) := f(x) + L \cdot \psi(x)$$

the following statements hold.

(continued)

3.9 Lipschitz Properties

(a) $\forall x \in X : \widehat{f}_L(x) \leq f(x)$.
(b) $\widehat{f}_L(\underline{x}) = f(\underline{x})$ and $\widehat{f}_L(\overline{x}) = f(\overline{x})$.
(c) $\max_{x \in X} (f(x) - \widehat{f}_L(x)) = (L/2)\, w(X)$.
(d) $\widehat{f}_L$ is monotonically decreasing on $[\underline{x}, m(X))$ and monotonically increasing on $(m(X), \overline{x}]$.
(e) $m(X)$ is a global minimal point of the function $\widehat{f}_L$ on X with minimal value $f(m(X)) - (L/2)\, w(X)$.

Proof For all $x \in [\underline{x}, m(X)]$ it holds

$$\psi(x) = m(X) - x - \frac{w(X)}{2} = \underline{x} - x \leq 0$$

and for all $x \in [m(X), \overline{x}]$

$$\psi(x) = x - m(X) - \frac{w(X)}{2} = x - \overline{x} \leq 0.$$

With $L > 0$ statement a follows. Statement b results from $\psi(\underline{x}) = \underline{x} - \underline{x} = 0$ and $\psi(\overline{x}) = \overline{x} - \overline{x} = 0$.

To see statement c, we note that for all $x \in X$

$$f(x) - \widehat{f}_L(x) = -L\psi(x)$$

holds, which implies

$$\max_{x \in X} (f(x) - \widehat{f}_L(x)) = -L \min_{x \in X} \psi(x).$$

Since ψ has slope -1 on $[\underline{x}, m(X))$ and slope $+1$ on $(m(X), \overline{x}]$, $m(X)$ is the unique minimal point of ψ with value $\psi(m(X)) = -(\overline{x} - \underline{x})/2$. This leads to

$$\max_{x \in X} (f(x) - \widehat{f}_L(x)) = -L\left(-\frac{\overline{x} - \underline{x}}{2}\right) = \frac{L}{2} w(X).$$

On $[\underline{x}, m(X))$ it also holds

$$\widehat{f}_L'(x) = f'(x) - L \leq |f'(x)| - L \leq \max_{x \in X} |f'(x)| - L \leq 0$$

and on $(m(X), \overline{x}]$

$$\widehat{f}_L'(x) = f'(x) + L \geq -|f'(x)| + L \geq -\max_{x \in X} |f'(x)| + L \geq 0,$$

which proves statement d. From this it follows that $m(X)$ is a global minimal point of $\widehat{f}_L$ on X, thus statement e. □

With the help of Theorem 3.9.10, one can specify a branch-and-bound procedure completely analogous to Algorithm 3.3 that terminates after a finite number of steps. The main advantages of this modification are that only once instead of twice continuous differentiability of f is required, and that, due to Theorem 3.9.10e, in line 9 no method from Sect. 2.8 is needed to calculate $\widehat{x}^\ell$ and $\widehat{v}^\ell$, $\ell = 1, 2$.

A significant disadvantage, however, is that the structure of f is poorly exploited, because for example minimal points $\widehat{x}^\ell$ of the relaxations at boundary points of sub-boxes X^ℓ cannot occur. Moreover, the bounds $\widehat{v}^\ell$ are usually worse than those from αBB-relaxations. On the one hand this increases the number of necessary iterations compared to αBB-relaxations, but due to the fast minimization of the relaxed problems, the CPU time per iteration is also significantly shorter. The predominate effect is problem-dependent. Finally, the generalization of the presented approach to the case $n > 1$ is not obvious.

Instead of mere Lipschitz continuity, there also exist more refined possibilities to exploit the slope information of the underlying function, for example *centric forms*, *Neumaier underestimators* and their combination, the *kites*. Details on this can be found for example in [29]. General introductions to the field of Lipschitz optimization are given in [20] and [21].

References

1. Adjiman, C.S., Dallwig, S., Floudas, C.A., Neumaier, A.: A global optimization method, αBB, for general twice-differentiable constrained NLPs – I: Theoretical advances. Comput. Chem. Eng. **22**, 1137–1158 (1998)
2. Adjiman, C.S., Androulakis, I.P., Floudas, C.A.: A global optimization method, αBB, for general twice-differentiable constrained NLPs – II: implementation and computational results. Comput. Chem. Eng. **22**, 1159–1179 (1998)
3. Alt, W.: Numerische Verfahren der Konvexen, Nichtglatten Optimierung. Teubner, Stuttgart (2004)
4. Bazaraa, M.S., Sherali, H.D., Shetty, C.M.: Nonlinear Programming. Wiley, New York (1993)
5. Boyd, S., Vandenberghe, L.: Convex Optimization. Cambridge University Press, Cambridge (2004)
6. Clarke, F.H.: Optimization and Nonsmooth Analysis. Society for Industrial and Applied Mathematics, Philadelphia (1990)
7. Dür, M.: A class of problems where dual bounds beat underestimation bounds. J. Global. Optim. **22**, 49–57 (2002)
8. Dür, M.: Copositive programming – a survey. In: Diehl, M., Glineur, F., Jarlebring, E., Michiels, W. (Eds.) Recent Advances in Optimization and its Applications in Engineering, pp. 3–20. Springer, Berlin (2010)
9. Fischer, G.: Lineare Algebra. SpringerSpektrum, Berlin (2014)
10. Floudas, C.A.: Deterministic Global Optimization. Kluwer, Dordrecht (2000)
11. Frank, M., Wolfe, P.: An algorithm for quadratic programming. Nav. Res. Logist. Q. **3**, 95–110 (1956)
12. Freund, R.W., Hoppe, R.H.W.: Stoer/Bulirsch: Numerical Mathematics 1. Springer, Berlin (2007)
13. Grant, M., Boyd, S., Ye, Y.: Disciplined Convex Programming. In: Liberti, L., Maculan, N. (Eds.) Global Optimization: From Theory to Implementation, pp. 155–210, Springer, New York (2006)
14. Güler, O.: Foundations of Optimization. Springer, Berlin (2010)
15. Hales, T.: A proof of the Kepler conjecture. Ann. Math. **162**, 1065–1185 (2005)
16. Hansen, E., Walster, G.W.: Global Optimization Using Interval Analysis. Marcel Dekker, New York (2004)
17. Heuser, H.: Lehrbuch der Analysis, Teil 2. SpringerVieweg, Wiesbaden (2008)
18. Heuser, H.: Lehrbuch der Analysis, Teil 1. SpringerVieweg, Wiesbaden (2009)
19. Hiriart-Urruty, J.B., Lemaréchal, C.: Fundamentals of Convex Analysis. Springer, Berlin (2001)
20. Horst, R., Pardalos, P.M. (Eds.): Handbook of Global Optimization. Springer, Boston (1995)
21. Horst, R., Tuy, H.: Global Optimization. Springer, Berlin (1996)
22. Jänich, K.: Lineare Algebra. Springer, Berlin (2008)

23. Jarre, F., Stoer, J.: Optimierung. Springer, Berlin (2004)
24. Jongen, H.T., Meer, K., Triesch, E.: Optimization Theory. Kluwer, Dordrecht (2004)
25. Kelley, J.E. Jr.: The cutting-plane method for solving convex programs. J. Soc. Ind. Appl. Math. **8**, 703–712 (1960)
26. Kirst, P., Stein, O., Steuermann, P.: Deterministic upper bounds for spatial branch-and-bound methods in global minimization with nonconvex constraints. TOP **23**, 591–616 (2015)
27. Lasserre, J.B.: On representations of the feasible set in convex optimization. Optim. Lett. **4**, 1–5 (2010)
28. Nesterov, Y., Nemirovski, A.: A general approach to polynomial-time algorithms design for convex programming, Tech. Report, Central Economic Mathematical Institute, USSR Academy of Sciences, Moscow, USSR (1988)
29. Neumaier, A.: Interval Methods for Systems of Equations. Cambridge University Press, Cambridge (1990)
30. Nickel, S., Rebennack, S., Stein, O., Waldmann, K.H.: Operations Research, 3rd edn. SpringerGabler, Berlin (2022)
31. Reemtsen, R.: Lineare Optimierung. Shaker, Maastricht (2001)
32. Rockafellar, R.T.: Convex Analysis. Princeton University Press, Princeton (1970)
33. Stein, O.: Twice differentiable characterizations of convexity notions for functions on full dimensional convex sets. Schedae Inf. **21**, 55–63 (2012)
34. Stein, O.: Grundzüge der Konvexen Analysis. SpringerSpektrum, Berlin (2021)
35. Stein, O.: Grundzüge der Parametrischen Optimierung. SpringerSpektrum, Berlin (2021)
36. Stein, O.: Grundzüge der Gemischt-ganzzahligen Optimierung. SpringerSpektrum, Berlin (2024)
37. Stein, O.: Basic Concepts of Nonlinear Optimization. Springer, Berlin (2024)
38. Vandenberghe, L., Boyd, S.: Semidefinite programming. SIAM Rev. **38**, 49–95 (1996)
39. Werner, J.: Numerische Mathematik II. Vieweg-Verlag, Braunschweig (1992)
40. Ziegler, G.: Lectures on Polytopes. Springer, New York (1995)

Index

active index, 76

barrier
　function, 102
　method, 102
　parameter, 103
bound, lower, 12
box, 123
　center, 132
　width, 132
branch and bound, 158, 164, 169

central path
　primal, 105
　primal-dual, 106
centric form, 178
chain rule, 43
cluster analysis, 10, 24, 26, 55
coercivity
　at ∞, 24
　on arbitrary sets, 30
complementarity, 75
composition, 126
cone, convex, 78
constraint qualification, 78
convex hull, 115
　function, 116
　problem, 118
convex relaxation
　of a function, 116
　of a set, 115
　of an optimization problem, 118
convexity
　hidden, 41
　of a function, 36

of a set, 36
of an optimization problem, 38
copositive, 108
critical point, 48
cutting plane method, 90
　Kelley's, 92

decision variable, 3
dependency effect, 125, 128
derivative
　first, 42
　second, 51
derivative matrix, 42
descent direction, 98
diameter, 133
disciplined convex programming, 109
distance, 17
duality
　gap, 63
　theorem, weak, 63, 65

epigraph, 9
　reformulation, 9, 33
　reformulation, generalized, 34
error bound, global, 176

factorizable, 126
feasible
　ε-minimal, 100
　dual, 64
　primal, 64
　set, 2
Fermat's rule, 48
Frank-Wolfe method, 98

function
  concave, 36
  convex, 36
  strictly concave, 36
  strictly convex, 36
  strongly concave, 36
  strongly convex, 36

Gaussian bracket, upper, 153
Gershgorin
  disk, 141
  interval, 143
gradient, 42
  method, 88

Hessian matrix, 51
Hoffman
  constant, 176
  lemma, 176
hyperplane, 4
  distance, 66

inconsistency, 16
infimum, 12
interior point methods, 102
interval
  n-dimensional, 123
  arithmetic, 120
  extension, 127
    inclusion isotonic, 130
    monotone, 130
    natural, 127
  hull, 124

Jacobian matrix, 42

Karush-Kuhn-Tucker point, 73

Lagrange
  dual, 62
  function, 59
level set, 20
linear independence constraint qualification, 79
Lipschitz
  constant, 170
  continuity, 170
log-likelihood function, 29

Mangasarian-Fromovitz constraint qualification, 79
maximal point
  global, 6
  local, 6
maximum likelihood estimator, 28, 30, 31, 49, 54, 56
mean value theorem, 44
minimal point
  ε-feasible, 96
  global, 6
  local, 6
minimal value, 6
Minkowski sum, 123
monotonicity
  abs, 27
  of a functional, 34
  of an operator, 57
multicriteria optimization, 6

Neumaier underestimators, 178
node selection, 161
norm-minimal solution, 5
normal vector, 4

objective function, 2
optimal
  point, 3
  value, 2
optimization problem
  convex, 38
  convexly described, 40
  copositive, 108
  linear, 41
  semidefinite, 108
outward rounding, 122

parallel projection, 15
parameter, 3
point cloud, 7, 9, 24, 26, 49, 54, 56
polyhedron, convex, 94
polytope, convex, 94
positive
  definite, 52
  part, 153
  semidefinite, 52
projection, 3, 17, 19, 23
  orthogonal, 4
  reformulation, 33

real numbers, extended, 13
redundant inequality, 112

separable, 32, 139
set
    bounded, 18
    closed, 17
    compact, 18
    convex, 36
    convexly described, 40
    feasible, 2
Slater
    constraint qualification, 79
    point, 79
spectrahedron, 107
stability number, 108
stationary point, 48
subdifferential, 44, 171
supremum, 12

Taylor
    model, 131
    theorem, 43, 52
tessellation, 149
trivial, 25

unboundedness, 16
unconstrained, 19
unsolvability, 14

variational formulation, 98

Weierstrass theorem, 19
    strengthened, 22
Wolfe dual, 64

SPRINGER NATURE

GPSR Compliance

The European Union's (EU) General Product Safety Regulation (GPSR) is a set of rules that requires consumer products to be safe and our obligations to ensure this.

If you have any concerns about our products, you can contact us on ProductSafety@springernature.com

In case Publisher is established outside the EU, the EU authorized representative is:

Springer Nature Customer Service Center GmbH
Europaplatz 3
69115 Heidelberg, Germany

The manufacturer's authorised representative in the EU is Springer Nature Customer Service Centre GmbH, Europaplatz 3, 69115 Heidelberg, Germany. If you have any concerns regarding our products, please contact ProductSafety@springernature.com

Printed and bound by CPI Group (UK) Ltd, Croydon, CR0 4YY

25/03/2026

02078172-0009